Artificial Intelligence

How Advanced Machine Learning will Shape the Future of Our World

© **Copyright 2018 - CHRISTINA AHMET - All rights reserved.**

The content contained within this book may not be reproduced, duplicated or transmitted without direct written permission from the author or the publisher.

Under no circumstances will any blame or legal responsibility be held against the publisher, or author, for any damages, reparation, or monetary loss due to the information contained within this book. Either directly or indirectly.

Legal Notice:

This book is copyright protected. This book is only for personal use. You cannot amend, distribute, sell, use, quote or paraphrase any part, or the content within this book, without the consent of the author or publisher.

Disclaimer Notice:

Please note the information contained within this document is for educational and entertainment purposes only. All effort has been executed to present accurate, up to date, and reliable, complete information. No warranties of any kind are declared or implied. Readers acknowledge that the author is not engaging in the rendering of legal, financial, medical or professional advice. The content within this book has been derived from various sources. Please consult a licensed professional before attempting any techniques outlined in this book.

By reading this document, the reader agrees that under no circumstances is the author responsible for any losses, direct or indirect, which are incurred as a result of the use of information contained within this document, including, but not limited to, — errors, omissions, or inaccuracies.

Table Of Contents

Table Of Contents

Chapter 1: Introduction

Chapter 2: What Is Artificial Intelligence?

Chapter 3: Machine Learning
Neural Network
The Resurgence in AI

Chapter 4: How Machines Learn
Supervised Learning
Unsupervised Learning
Reinforcement Learning
Genetic Breeding Model (Or Evolutionary Model)
Deep Learning and Recursive Neural Networks

Chapter 5: AI and the Internet of Things
The Internet of Things
Why Share Data?
Privacy Implications
What Is Needed?
Intelligent IoT
Minimizing Unplanned Downtime
Operational Efficiency
Newer Products and Services
Greater Risk Management
Implications for Enterprises

The Future of IoT is AI

Chapter 6: Opportunities for Artificial Intelligence

Chapter 7: Threats of Artificial Intelligence
General-Purpose AI
White-Collar AI
Creative AI

Chapter 8: How AI is Used in Different Industries
A Great Help for Humans
Finance and Banking
Heavy Industries
Healthcare
Retail
Technology
Higher Education
Energy and Utilities
Transport

Chapter 9: How Humans and AI can Work Together
Where Humans Are Better than Bots, and Vice Versa
What AI Does Best
What Human Does Best
Verdict
Humans Helping Machines
Training
Explaining
Maintenance

Machines Helping Humans
Amplifying
Interacting
Embodying

Chapter 10: Ethical Challenges of AI
Ethical Issues
Unemployment
Inequality
Humanity
Biases in Algorithms
Transparency of Algorithms
Supremacy of Algorithms
Fake News and Fake Videos
Weapons
Self-Driving Cars
Rights of the Machines
Privacy and Surveillance
Access to AI Technology
Error Prevention
Singularity
Ethical AI
Challenges of AI
What AI did Wrong in the Past
What Have Been Done to Make AI more Ethical
State's Involvement
Final Thoughts

Chapter 11: Conclusion

Chapter 1: Introduction

In this digital age, many people are surrounded by modern technologies. They have access to a world with a device as small as their palm. A few hundred years back, it would be considered to be witchcraft when one was able to talk to someone halfway across the globe, see the Great Wall of China while still being in the United States, or participate in a meeting from the living room. Now, these superpowers do not seem like they are worth much anymore. Technological advancements have brought humanity very far. These advancements are intended to make everyday tasks easier. Surgeons initially learned how to perform surgery, until the machine was invented to help them operate with greater accuracy. A long time ago, a man on a horse was the best internet connection one could ask for. Now, text messaging is instant. Humanity is a race that can stand and walk on their two feet as the sole means of transportation until they invented a way not to do so in the form of horses (eventually replaced by cars) or even planes. It may not sound like much, but try to explain to the people in the medieval age how a plane, which is basically a big metal tube with wings, is able to carry tens of passengers and travel around the globe in less than 3 days.

It is undeniable that technology has come a very long way indeed. However, it comes at a price. Improper use of medicine can lead to the evolution of diseases known as superbugs. Certain substances are abused. Communication, while enabling a teenager to talk to his long-distance girlfriend thousands of miles away, turns him cold and distant to his family who are sitting with him at the dinner table. As previously mentioned, horses were initially thought to be the main transportation solution but became obsolete with automobiles. Factory workers, fishermen, and farmers are slowly being replaced with machines that can do a lot more work, costing less time and even money. Therefore, it is equally undeniable that technological advancements change society as well.

Amongst those advancements, concerns of artificial intelligence are raised. Some old movies may inspire ideas about the machines turning against their creators, and such references can be heard tossed about in any discussion about AI. On a more serious note, some people saw what technology did to horses. They wonder if they will be next. Worryingly enough, physicist Stephen Hawking as well as Elon Musk, the CEO of Tesla Motors, warned about the development of artificial intelligence.

According to Stephen Hawking, the emergence of artificial intelligence could very well be the worse event in the history of civilization unless society finds a way to control its development. One of his comments stressed the fact that theoretically, computers can emulate human intelligence and eventually exceed it.

Currently, artificial intelligence lives amongst the human population. They reside in smartphones. They help people find content on the internet. They learn the behavior of their owners and put out relevant, interesting content to enhance their owner's experience while they are browsing on the internet.

When artificial intelligence is so deeply entrenched like this, concerns are justified. While the future remains uncertain, the present is clear as crystal. In this book, everything about artificial intelligence will be discussed in detail, including its history, how it works, and its effect on society as a whole.

Chapter 2: What Is Artificial Intelligence?

The word AI itself consists of two words: artificial and intelligence.

Here, artificial is something that is not real, simulated, but not entirely fake regarding being a scam. Think of the grass used to cover a stadium ground. That is artificial grass, but it sort of function similarly to real grass. It is widely used in sports because it is a lot more resistant and easier to care for. The point is, artificial means something that could replace genuine items because the former has better qualities in a certain context.

Intelligence is a very complex term. One can define it in various ways such as logic, self-awareness, learning, emotional knowledge, planning, conscience, creativity. The list goes on. Human is the most intelligent species mainly because we have certain mental capacities that some animals lack. Humans can perceive their environment, learn from it, and take action based on what they discover. Animals, of course, are capable of doing just as much, to a certain degree. However, the human brain evolved exponentially until they invented languages, tools, and eventually became the dominant species. Both humans and animals possess what is known as natural intelligence.

There have been some arguments about the existence of plant intelligence. Plants show their intelligence differently than humans or animals. Plants do not have brains or any neural network to speak of, but they are fully capable of reacting to their environment. There was an experiment in which a plant was placed inside a dark box with only a small hole through which sunlight could shine. As the plant grows, it leans toward the sunlight, which is a clear indication that they can react to their environment, although they do not have a nervous system. Plant intelligence is an interesting topic of its own mainly because it is not as easily perceived as that in humans or animals.

Nowadays, a new type of intelligence pops up, known as artificial intelligence. Broadly speaking, artificial intelligence (or AI for short) is, as the name suggests, a simulation of human intelligence process by machines such as in a computer system. AI systems generally have some of the behaviors associated with human intelligence such as planning, reasoning, problem-solving, knowledge representation, learning, motion, perception, manipulation, and to a certain extent, social intelligence and creativity. There are two main types of AI, narrow AI, and general AI.

Narrow AI exists in computers nowadays. These AI are taught or learn themselves how to carry out specific tasks without being programmed explicitly how to do so. A perfect example is the speech and language recognition of Siri, the virtual assistant in iPhones, vision-recognition systems of self-driving cars, or even the hidden AI that show you the advertisement on websites that are relevant to the viewer. These AI are called narrow AI because they can only learn or be taught how to do a specific task. Narrow AI can do many things ranging from interpreting video feeds from drones that are conducting visual inspections of infrastructures to mundane tasks such as organizing personal and business calendars. They can respond to simple customer-service queries, coordinating with other intelligence systems to book a hotel room at just the right price and location. It can even help radiologists spot potential cancer tumors in X-rays, detect wear and tear in elevators, and flag inappropriate content online to protect the viewers.

General AI, on the other hand, is an entirely different entity altogether. Unlike narrow AI that can only learn one thing, general AI has the same adaptable intelligence found in humans. This flexibility allows them to learn how to do many different tasks, such as building spreadsheets, haircutting, or even driving based on its accumulated experience. This sort of AI is the sort found in famous movies like Terminator. So far, it does not exist yet, and AI experts are divided over when it will be a reality.

According to a survey conducted in 2013 by AI researchers, Vincent C Müllet and philosopher Nick Bostrom among four groups of experts reported a 50% chance that Artificial General Intelligence (AGI) would be developed around 2040 or 2050. The chances are higher, up to 90% by 2075. The emergence of a superintelligence is also predicted in the survey, which Bostrom defined as an intellect that exceeds the cognitive performance of humans by a long shot in all domains of interest. This was expected about 30 years after the development of AGI.

Still, as of now, some AI experts remain skeptical

about this projection because the human brain is not yet fully understood, so it is theoretically impossible to develop an intelligence that imitates the human brain. It is also possible that AGI is still hundreds of years away from being developed.

Back in 2017, a demonstration of what AI could do came in the form of a killer AI from Elon Musk-backed startup, OpenAI. This artificial intelligence was able to topple the world's best Dota 2 players.

Dota 2 is an incredibly complex game in which two teams of 5 players compete to lay waste to the opposing team's base. Given the fact that there are more than 100 playable characters with unique abilities to choose from, not to mention dozens of items and item combinations, it is virtually impossible to comprehend the complexity of this game. Given its complexity and competitiveness, it has become one of the most famous e-sports in the world with yearly Dota 2 tournaments by Valve.

In the yearly tournament in 2017, a surprise segment introduced the best new player in the world, which happened to be a bot from OpenAI. According to the engineers from the nonprofit, the bot learned enough to beat Dota 2 professional players in just 2 weeks of real-time learning. During that time, the bot amassed a lifetime of experience by using a neural network, running several instances of the bot simultaneously, and speeding up the game simulation to hasten the learning process. The bot started with only a few instructions such as dying is bad, doing damage to the enemy is good, etc. Then, it eventually learned to explore its surroundings, learn the inner workings in the game, and compete against professional human players with a high degree of success. Just for comparison, the bot accumulated enough experience equivalent to professional players in two weeks whereas the players take several years. This demonstrates the astonishing and frightening capability of AI through machine learning.

Chapter 3: Machine Learning

Machine learning is a process in which a bot accumulates a large amount of data, which it uses to learn how to perform a certain task, such as recognizing speech or captioning a photograph. The key to the process of machine learning is the neural network.

Neural Network

This network is inspired by the brain because both have some similarities. In the neural network, there are several interconnected layers of algorithms called neurons. These algorithms feed data to each other, and they can be trained to carry out specific tasks by giving simple instructions highlighting important attributes to input data when it passes between layers.

It may sound complicated, and it is. Basically, the neural network functions just like a human brain. Each neuron does a specific thing, and it is connected to the adjacent neuron. It is possible to study what a specific neuron in a human brain does, and the general purpose of a cluster of neurons can be understood, but the whole human brain remains a mystery. The same could be said for AI neuron network. A single algorithm can be studied, and the purpose of a cluster can be understood, but the whole system behind the AI cannot be deciphered. Still, one thing remains very clear: the entire system just works.

There is also a subset of machine learning known as deep learning. Here, neural networks are expanded into many sprawling networks in a massive number of layers that are designed to interpret a lot of data. These deep neural networks fueled the leap forward in the ability of computers to perform tasks such as recognizing speech and computer vision.

There are also many types of neural networks, and all of them have their own strengths and weaknesses. Recurrent neural networks, for example, are a type of neural net that is well-suited to language processing and speech recognition. Convolutional neural networks, on the other hand, are catered toward recognizing images. The design of these neural networks is also evolving. Researchers are refining a more effective form of deep neural network known as long short-term memory (LTSM) to allow the networks to operate quickly enough to meet the demand of systems such as Google Translate.

Another area of AI research is known as evolutionary computation or the evolutionary model. This model uses Darwin's famous theory of natural selection – survival of the fittest. Here, the genetic algorithms undergo random mutations as well as combinations between generations in an attempt to create the right solution to a problem (more to be discussed below).

The Resurgence in AI

Perhaps one of the biggest breakthroughs for the research in AI technology is in the field of machine-learning, especially in the field of deep learning. Of course, its success comes with the fact that a wide variety of data can be gathered relatively easily, not to mention the explosion in parallel computing power in recent years. During that time, the use of GPU clusters to train machine-learning systems has become even more widespread.

These clusters not only offer a lot more powerful systems to train machine-learning models, but they are also available widely as cloud services over the internet. Over time, major tech firms such as Google and Microsoft have moved to utilize specialized chips that are tailored to running and training machine-learning models. The best example of these custom chips is Tensor Processing Unit from Google. Its latest version allows the acceleration of the speed that machine-learning models study information from data, not to mention the speed of how bots are trained. These chips are not simply used to train up models for DeepMine and Google Brain. They are also used to train the models that underpin the famous Google Translate and image recognition of Google Photos, not to mention all the other services that allow the public to build machine-learning models on their own using Google's TensorFlow Research Cloud. The second generation of these customized chips was unveiled at Google's I/O conference just last year, promising even greater speed for Google machine-learning model that is used for translation. Here, it is said that the bot can translate in half the time it would take a large bunch of high-end graphics processing units.

Chapter 4: How Machines Learn

Previously, the theories behind machine learning were explained thoroughly. However, to better understand the entire process, it is necessary to illustrate it in real-life events.

On the internet, the algorithms are everywhere. The videos people watch are brought to them by the algorithms. When they click on the video, the algorithms take note. When people go on Facebook, the algorithms decide what people see. When people buy something, they set the price, communicate with their fellow algorithms at the bank to monitor the transaction. The stock market is full of bots trading with each other instantaneously. In fact, this book could be brought to the users by algorithms as well.

Before, humans built algorithmic bots by providing them with instructions that humans can easily explain. If A, then B. However, one could easily see the limitations in manual programming. Certain problems are so vast that it is virtually impossible for humans to program an effective bot. Looking back at the stock market, there are millions of transactions going on every second. How does a person program a bot to know which one is bad? There are countless videos on YouTube. How is it possible to program a bot that can recommend the right video for the viewer fairly consistently? How does one program a bot to determine how much a customer is willing to pay for that airline seat? It is true that bots may not give perfect answers, but at least they do a better job than humans.

Funnily enough, while they can work wonders, no one really knows how they work. Not even their creators know how they function. Even so, many companies are very protective when it comes to how their bots work because they are incredibly valuable employees. How the bots are built is a secret. Whatever their models may be, there are three main categories for machine-learning, which are supervised, unsupervised, and reinforcement learning.

Supervised Learning

A common method for teaching an AI system is

by using a large chunk of data and labeled examples. Here, the bots are fed a huge amount of data and then they need to study the data and highlight the features of interest. If the data fed are photos and the bots need to say whether they have a dog or written words, then they will look through the photos and try their very best to understand. Again, neither the human programmer nor the bots themselves know how they think, but at least they work relatively well for a robot. When these bots are trained, which means that they can accurately perform their tasks using the labeled examples, then they should be capable enough of doing the same thing when they are fed new data.

This process of teaching a bot using examples is known as supervised learning. Normally, the task of creating and labeling examples fall into the area of responsibilities by online workers who are generally employed through platforms such as Amazon Mechanical Turk.

Of course, to create a bot that can reliably tell the difference between a cat and a catfish would require a lot of labeled datasets, which can mean millions of images and examples just to teach the bot to do one thing right. Still, in an age of big data and widespread data mining, data is relatively easier to access than it once was a decade ago. There are also training datasets available in huge quantities, and they are still growing, such as Google's Open Image Dataset, which contains about nine million images. At the same time, YouTube has its own labeled video repository known as YouTube-8M which has the link to over seven million labeled videos. ImageNet, which is one of the earliest databases invaluable to machine-learning, now has more than 14 million categorized images. As such, a company or those who want to design bots might give value to computing power more than having access to the labeled database. In recent years, Generative Adversarial Networks have demonstrated how machine-learning systems, when fed only a small amount of labeled data, can create its own fresh data with which to teach themselves. This method could be tremendously useful and can potentially lead to the rise of semi-supervised learning where bots can learn how to perform a certain task using a significantly smaller amount of labeled data than what is originally needed to train a bot using a supervised learning model.

Unsupervised Learning

On the other hand, unsupervised learning does not need as much human involvement. Here, the algorithms try to find patterns in data and look for similarities that can be used to categorize that data. A perfect example here is by clustering fruits that weigh or look similar or cars with a similar weight or size.

Here, the algorithms are not created in advance to look out for a particular set of data. It just takes a look at the data it is given and tries to group everything in there by its similarities. A perfect example of this is Google News, which groups stories on similar topics on a daily basis.

Reinforcement Learning

The best way to describe reinforcement learning is rewarding a pet with a treat when it successfully performs a trick. Except, the pet is a bot, and the reward is the desired outcome which so happens to also be the trick.

Here, the system tries to maximize a reward based on its input data and continues to learn until the best possible outcome is achieved.

A good example here is by looking at Google DeepMind's Deep Q-network. It has been used to defeat humans in many classic video games. Here, the bot is given a massive amount of data, from the pixels for each and every game and it attempts to figure out different information such as how to jump so that the character does not fall into the pit. The bot also looks at the score in each game and figures out a way to maximize the score in different circumstances. In the case of Mario Bros., the bot will chart a route to guide Mario through each level to collect all the coins, defeat all the monsters, while doing so in the least amount of time possible.

What these bots can achieve is miraculous, and it seems that the sky is the limit. As long as there is data, enough computing power, and some time, a bot can be taught just about anything. But how do they learn exactly? By studying how such a bot can be built, it is possible to understand how they learn without getting into excruciating details of how each and every neuron works.

Suppose that a bot is needed to sort through a collection of photos and divide them into two categories: three or trees. It is easy for humans, even children, but it is impossible for the human to program a bot to recognize the difference.

Humans can describe a tree by saying colors, shape, or material, and a number by its shape. But those are words, and bots cannot understand words. They read and understand algorithms, which is impossible to program. So, how does one build a bot to do this job? Instead of building a bot that can sort through the photos, a bot that builds other bots and a bot that teaches those bots are created. The brain of these bots is vastly simpler, which is something a talented programmer can achieve. Here is where the path of machine-learning really branches out. There are two main models of how to make machines learn: genetic breeding models, and deep learning and recursive neural networks.

Genetic Breeding Model (Or Evolutionary Model)

The builder bot builds bots almost by random, except for some instructions given by the programmer. It just connects the wires and modules in the student bots' brains, resulting in wildly random bots being sent to the teacher bots.

Of course, the teacher bot cannot tell the difference between a three or a tree either (otherwise, there is no need for a builder bot and a teacher bot). The teacher bot does not teach the student bots. The programmer gives the teacher bots some photos of trees and threes as well as an answer key telling which photo is which. Here, the teacher bot tests the student bots. Because the student bots are initially built by random, they will perform poorly at the test. Then, the results are sent to the builder bot. The best student bots are kept, and the rest are discarded. Then, the builder bot gets back to work by creating copies of the student bots and adding additional changes. Then, the second generation student bots are sent to the teacher bot. The teacher bot hands out the test, corrects them, send the result back, and the builder bot gets back to work again. This cycle continues indefinitely.

Basically, this is the "spray and pray" method of programming, which should not work, yet does. This is partly because, in every generation, the builder bot only keeps the best student and then discards the rest. Moreover, there is not just one student bot. There are countless, and the teacher bot can also have millions of questions as well. This building, testing, discarding cycle continues until the desired result is achieved.

At first, the best bots are just lucky because they have the right algorithms. However, when one combines enough lucky bots and keeps only what works, and then randomly makes a few tweaks, there will be a time when a student bot can do its job reasonably well. While the student bots are copied and changed, the average test score is also raised to improve the bots even further. Keep this up, and a bot will eventually emerge that can tell the difference with great accuracy. But how the student bot does this, nobody knows for sure. Even the student bot itself cannot understand it either. Still, it just works.

However, the bot still has a narrow AI. In this example, the student bot is good at exactly the only kinds of questions it has been taught to – telling the difference between a tree and a three. It can work with photos, but it is useless with videos. It will fail miserably if the photos are upside down. There are instances where it makes blatant mistakes as well. Suppose that a child dresses up like a pine tree. Humans can tell that it is obviously a child, but the bot maintains that it is a tree. So, how does a programmer compensate for this error? More test materials. For the previous example, that means more pictures. That is why many companies are obsessed with collecting data. The more data they have, the longer the tests are. The longer the tests, the better the bots become. There is also CAPTCHA (Completely Automated Public Turing test to tell Computers and Humans Apart) tests on some websites. The purposes of these tests are twofold. It is there to prevent bots from entering the website. These tests are also designed to make the website visitors help build the test for the bots so they can read, count, or tell cliffs from roads. Many CAPTCHA tests nowadays concentrate on identifying road, vehicles, signs, storefronts. Considering that the next technological race in the automobile industry is a self-driving car, ever wondered what could that be a building test for? The CAPTCHA tests previously discussed require humans to make correct enough tests. However, there is also another kind of test that makes itself. It is a test on humans.

To illustrate, suppose that a video website called NetTube wants to keep its users on their website as long as possible. How could that be achieved with bots? Since it is easy to measure how long a user stays on the website, the teacher bot simply gives each student bot a bunch of NetTube to oversee. The student bots are then tasked to observe the behavior of its humans, look at the videos the humans watched, and try to pick the next video that might just be enticing enough to keep the user online. The longer the average online time is, the higher the bot's test score is. The cycle of build and test continues and roughly a million cycles later, there's a student bot that is pretty good at keeping the users watching, at least compared to what a human could build. Again, when people ask how this bot selects videos or what algorithm goes inside the mind of this intelligence, there is not a great answer other than pointing out the fact that the bot has access to the user data and how the programmer or human overseer directs the teacher bot to score the test. But what goes through the mind of the bot is a mystery. All that is known is that the bot gets to live simply because it is slightly better than others until a better bot comes along.

Deep Learning and Recursive Neural Networks

While the genetic breeding model of how to make a machine learn using a genetic code that is an older code (that still works), the current trend is deep learning and recursive neural networks. Here is where the algebra going on inside the mind of the bot becomes even more complicated.

Deep training or deep learning is different from the genetic breeding model or evolutionary model. Here, this technology enables the bot to generalize data. For instance, a computer that is trained by a neural network cannot recognize handwritten letters well. So, to allow the bot the capability of handwritten text analysis and recognition, one must provide a large set of data with every handwritten letter in it. That way, the bot is able to recognize handwriting. Again, the larger the sample, the more accurate the recognition becomes.

Deep training got its name by how it analyzes the data it is given. It does so by analyzing each element or layer based on their priorities. For example, suppose that a company wants to create a bot that can tell the difference between a boy and a girl. The programmer simply hands the bot some pictures of boys and girls. Because there are multiple layers, the bots will have multiple neurons and triggers. To simplify, on the first layer, which can be said to be the first step of analysis, the bot looks at simple visual triggers such as brightness differences. Then, on the second layer, or the second step of analysis, the bot looks at more complex triggers, elements, or variables such as corners and circles. Going to the third step of analysis, or the third layer, the bot observes the details of the human faces. Here, the complexity of layers increases almost exponentially. Of course, the network decides which elements in the picture it should prioritize first. Then, it ranks all of those elements in order of importance so to optimize the recognition process and better understand what is in the photo. So, what is the difference between machine learning that utilizes the evolutionary model and deep training?

To put it simply, both have the same builder and teacher bots. The difference is that deep training does not have an infinite warehouse packed with an infinite amount of student bots. There is just one student bot. The teacher bot has the same test, but the builder bot is not building new bots. His job is to "fine-tune" the student bot. Think of it not as a builder bot, but as a dial adjustment bot. The dial controls how sensitive a connection in the student bot's head is. Think of the old dials on radios in cars. While the driver might not know the exact frequency, he can tell if he is getting closer. Using that guide alone, he can eventually adjust the dial until he gets to the radio station he wants to listen to. Because there are a lot of connections in the student bot's head, there are a lot of dials to adjust. The dial here represents the layers discussed previously. The teacher bot shows the student bot a photo and the dial adjustment bot adjusts the dial little by little, on a scale of stronger to weaker, as the student bot gets closer to the answer, and continues to do so until it answers the question correctly. Deep training basically works this way, but there are hundreds of thousands of dials and a lot of math. Here, it is worth noting that these adjustments are just for one question. The teacher bot then moves on to the next question (using the previous example, another photo) and the dial adjustment bot readjusts the dials once again so the student bot can answer both questions correctly. As the test gets progressively longer, the adjustments become exponentially complex as the dial adjustment bot continue to fine-tune the student bot. Of course, the student bot can and will get better the longer it trains, but it still suffers from limitation discussed previously for bots that learn using the evolutionary model.

For example, suppose the bot is created to tell a boy from a girl. Because it has multiple layers, it will look for different things with different priorities to determine if the picture contains a picture of a boy or a girl. Suppose that the bot looks for and prioritize the bunch of pixels that would be classified as hair in the first layer. In the second layer, it defines the clothing. The third, it looks for the shape of the face, lips, and so on. Therefore, when presented with a picture, the bot will first look at the hair. Is it long or short? What kind of hairstyle? Then, what is the person wearing? Next, what does the person in the picture look like? As the layer goes deeper, everything becomes even more complex. Using this example, a possible trick can be used to trick the bot. Using the magic of makeup, a girl can transform into a boy, which can quite easily fool the bot unless it underwent additional vigorous training in recognizing even more features. Here, the use of neural networks and deep training allows the AI to be more effective when it needs to take on a complex job, unlike the evolutionary model which is mostly based on random chances. So, are there any actual developments that utilize deep learning?

Many projects that use deep learning are mostly used in photo or audio identification, not to mention the diagnosis of diseases. It is used in Google Translate, which allows for instant translation by simply taking a photo of the text. In this case, deep learning gives the bot the ability to recognize texts in pictures and translate them. DeepFace is a system that works with photos for face recognition. DeepFace is able to recognize human faces at about 97.25% accuracy, which is almost as good as any real human.

Back in 2016, Google released WaveNet. It is a system that can simulate human speech. It is possible by uploading millions of minutes of recorded voice requests to the system. The data has been used in a Google project called "OK Google." Then, the neural network went through all those voices and put together sentences that just sound natural, stresses, accent and all, without illogical pauses.

Deep learning can also split images or videos into different segments semantically. The AI is fully capable of seeing when there is an object in the picture and outline its frame with remarkable accuracy. This technology is utilized in self-driving vehicles to determine whether obstacles are marked on the road. Additionally, it can read information from different traffic signs to avoid accidents. Moreover, neural networks are also used in medicine, such as one that determines diabetic retinopathy just by using photos of the patients' eyes. The result is so great the US Department of Health has already allowed the use of this technology in public clinics.

Chapter 5: AI and the Internet of Things

It is not a mystery that the Internet of Things (IoT) is getting smarter. Many companies across the globe are incorporating artificial intelligence with machine-learning technology into their IoT applications. What this AI does best is find insights in data that would otherwise take too long or be outright impossible for humans. However, before understanding how AI could improve IoT, it is best to understand just what IoT is.

The Internet of Things

Think of smart toasters or connected rectal thermometers or fitness collars for animals. These are just mundane, silly things that are connected to the web. These devices are a part of the Internet of Things.

Machines or devices that are connected to the

web allow the potential for a fourth industrial revolution. Many experts predict that more than half of new businesses will be based on the IoT by 2020.

The Internet of Thing is often simply described as all the devices that are connected to the internet. Recently, however, the same term is also used for objects that communicate with each other. To put it simply, the IoT is comprised of millions of devices worldwide ranging from sensors to smartphones that are all interconnected.

When all these connected devices are combined with automated systems, it is actually possible to gather information, analyze all the data collected, and create an action plan to help someone accomplish a certain task. In the IoT, three major elements are identified: devices, the data they store, and the networks which those devices use to communicate. IoT allows devices that are on a closed private internet connection to talk to each other as well as bringing all those groups of devices in the private network to communicate with other devices across different networking types to create a more interconnected world.

Why Share Data?

There is a comedic saying that some scientists are so preoccupied with: whether they can do something, without even stopping to think whether they should. The same could be said for certain smart devices. There is a smart salt shaker with built-in microphone that can dispense just the right amount of salt, a smart trash can that can (pun inevitable, apologies) keep track of its user's waste, a smart toaster that can send and receive text messages, an egg tracker, or a smart mirror that gives information about the weather. Just because a device can be connected to the internet, it does not mean that it should. Still, each device that does so, one way or another, collects data with a specific goal that can be useful for the buyer and ultimately impact the wider economy.

Looking at their industrial applications, sensors on product lines alone increase production efficiency and reduce waste. A study has given an estimation of 35% of US manufacturers are using data from smart sensors inside their set-ups.

IoT allows everything to be more efficient, ranging from mass production or major tasks to small, mundane tasks. No matter how big or complicated the job may be, IoT allows humans to be more efficient in what they do, thus allowing them to do more things in the same amount of time. Moreover, the quality and scope of the data in IoT creates an opportunity for a much more contextualized and responsive interaction with devices that can create a potential for change.

Privacy Implications

Of course, cybersecurity concerns have never been greater than nowadays. It goes without saying that everything that is connected to the internet can always be hacked. IoT products are no exceptions. A perfect example is a scandal surrounding VTech losing its videos and pictures of children using its connected devices, which is caused by an insecure IoT system. To have a better grasp of the situation, VTech's electronic toys handled a hack attack very poorly. In 2015, more than 6 million children's accounts were compromised from the breach, which happened to allow the hacker access to photos and chat logs in VTech's toys.

Certain devices have surveillance capability. This is yet another problem because those devices can also be connected to IoT, which allows the data to be siphoned off to those who need it. This ultimately gives a third party the ability to observe users. For example, a connected fridge can track food usage and consumption which tells users when they should buy groceries. This is convenient, but many people may overlook a hidden feature. The data from the fridge can also be used by takeaway restaurants to set the price just right when that individual does come by when their fridge is empty.

According to James Clapper, the US director for national intelligence, intelligence services might utilize IoT for identification, surveillance, location tracking, monitoring, targeting for recruitment, or even to gain access to networks or user credentials.

What Is Needed?

Compatible standards are the significant issues lying at the center of the establishment of a vast, reliable IoT network. It has been established that devices do not speak or understand English as humans do. Therefore, a compatible means of communication is needed so that all devices can communicate amongst themselves to effectively transfer or share the data they are recording. If they run on different standards, there will be incompatibilities which could lead to miscommunication or lack of it. Therefore, the need for standardization is on the rise.

Microsoft has introduced its own system for IoT devices in an attempt to tackle this issue on an enterprise scale. This system is called IoT Central, which gives businesses a central platform that is managed for setting up IoT devices.

Presently, IoT will impact anything with a high cost of not intervening. IoT will generally impact mundane, daily issues such as locating a parking space or confirming whether there is milk in the fridge. However, with the introduction and development of artificial intelligence, the capacity of IoT may increase tremendously.

Intelligent IoT

Artificial intelligence has made IoT even smarter. The former, especially ones that use machine-learning, is incorporated into the applications of the latter to improve its capacity regarding operational efficiency and downtime reduction. The key here is to find insights in data. With a wave of investment, a flood of new products, and a rising trend of enterprise deployment, AI is making its appearance known to the world and how it could work in unison with IoT. Because of that, companies all over the world are eyeing AI when they want to get more value from their existing IoT deployment.

It is clear that Ai has a role in IoT applications and deployments because of the apparent shift in the behaviors of companies in this area. There is a sharp increase for venture capital investments in IoT start-ups that utilize AI. Several companies have acquired dozens of firms that work to integrate AI into IoT for a few years now. Plus, major vendors of IoT platform software now offer AI capabilities that are integrated into the IoT that utilizes machine-learning-based analytics.

Because AI can produce insights from a large amount of data, machine-learning technology allows AI to go through millions of pieces of data and connect the dots when it finds any connections. All the data is collected by all the devices as part of the IoT such as temperature, pressure, humidity, air quality, sound, and vibration, are all generated by sensors and devices which AI could use to interpret. Many companies also find that machine-learning can have a great advantage over traditional business intelligence tools when it comes to the analysis of IoT data. Initially, data collected by IoT devices must be viewed and interpreted manually. Now, the application of artificial intelligence allows the interpretation to be more efficient by making operational predictions up to 20 times faster, not to mention that AI can deliver predictions that are much more accurate than threshold-based monitoring systems. Speech recognition and computer vision can also assist in extracting additional insights and information from the data that IoT collects, which initially had to be reviewed by humans.

This IoT-AI duo has proven itself to be tremendously helpful to companies, especially when it comes to minimizing unplanned downtime, operational efficiency, newer products and services, and better risk management.

Minimizing Unplanned Downtime

Unplanned downtime from equipment failure disrupts the processes in many sectors that not only costs them time but also a lot of money. For example, one study has shown that offshore oil and gas operators suffer an average annual loss of $38 million. Another source estimated that unplanned downtime costs $50 billion annually for industrial manufacturing in total and equipment breakdown is the cause for 42% of it. So, how could IoT and AI be used in this context to minimize downtime? Through predictive maintenance.

Predictive maintenance uses analytics to predict equipment failure before it happens so to allow humans to plan for downtime to arrange orderly maintenance procedures. This can mitigate the damaging economics of unplanned downtime. For manufacturing industries, Deloitte, a multinational professional services network, discovers that this predictive maintenance technology can cut down the time needed to plan maintenance by up to 50%, reduce overall maintenance costs by up to 10%, not to mention that equipment's uptime and availability are also increased by up to 20%.

Since AI technologies, especially machine-learning, can assist in the identification of patterns and anomalies and establish predictions based on a massive amount of data, they have proven themselves to be very handy in implementing predictive maintenance. SK Innovation, a South Korean oil refiner, expects to save billions just by using machine-learning to predict connected compressor failure ahead of time. Similarly, French power utility EDF Group was saved from over a million dollars' worth of expense just by using AI to predict equipment failure. Meanwhile, Italian train operator Trenitalia also expects to avoid unplanned downtime and save around 10% of its 1.3 billion euros in annual maintenance expenses.

Operational Efficiency

IoT powered by AI can also accomplish more than just prevent unplanned downtime. It can also help improve operational efficiency. This is because machine-learning can also generate quick and accurate predictions as well as deep insights, not to mention that it also gives AI technologies the ability to automate a growing variety of task.

To illustrate, for Hershey, which is known for its chocolate products, they need to control the weight in their products. This process is critical because every 1% improvement in weight precision can result in more than $500,000 savings for a 14,000-gallon batch of products such as Twizzlers. Hershey used IoT and machine-learning to reduce weight variability during the production process significantly. The data from IoT is recorded and analyzed every second, and prediction of weight variability can be used by machine-learning models, which allows for up to 240 process adjustments daily. Before the machine-learning-powered IoT solution was installed, there was only a 12 process adjustment.

Predictions by AI also helps Google by reducing up to 40% of the expenses from the cooling of its data center. Here, the solution is utilizing the data from sensors in the facility to predict temperature and pressure for the next hour. That way, the power can be accurately dedicated to optimizing the cooling process.

Machine-learning has also helped a shipping fleet operator in producing insights that help the operator in taking a, what initially seemed counterintuitive, action that actually saved them a lot of money. Here, the data collected from sensors aboard the ship was used to identify the correlation between the expenses on cleaning the ships' hulls and fuel efficiency. According to that analysis, cleaning the ships hulls twice every year rather than once every two years would allow them to save up to $400,000 in fuel expenses. It is true that the cleaning expenses will be quadrupled, but the increase in fuel efficiency far outweighs the expenses.

Newer Products and Services

AI technology coupled with IoT can create the foundation upon which new and improved products can be created. For example, for the inspection services that are based on robots and drones from General Electronics, the company is looking to AI to assist in the automation of both navigations of inspection devices as well as identification of defects from the data from the robots and drones. That means safer and more accurate inspection, not to mention that the processes become 25% cheaper for the client. In the healthcare industry, Thomas Jefferson University Hospital in Philadelphia also seeks to improve the experience of patients using natural language processing that allow the patients to control the room environment and request various information with just voice commands instead of asking for a nurse.

Rolls-Royce, a famous car brand company, also aims to introduce a new offering featuring airplane engine maintenance services that are supported by IoT. The company has plans to utilize machine-learning to spot patterns and identify operational insights that will be sold to airlines. Navistar, an automotive manufacturer, also looks to machine-learning analysis of real-time connected vehicle data to allow for a new stream of revenue in diagnostics as well as predictive maintenance services. According to Cloudera, Navistar's technology partner, these services have helped reduce downtime for almost 300,000 vehicles by up to 40%.

Greater Risk Management

Organizations can better understand and predict a variety of risks as well as automate for rapid responses thanks to some applications that pair IoT with AI. Here, they enable those organizations to manage their worker safety, financial loss, and cyber threats better.

For example, Fujitsu, a Japanese multinational information technology equipment and services company, has piloted the use of machine-learning. It is used to analyze the data collected from connected wearable devices to accurately estimate the factory workers potentially threatening heat stress that is accumulated over a period of time. Many banks in India and North America have also started evaluating AI-based real-time identification of suspicious activities from numerous connected surveillance cameras at ATMs. In the insurance industry, vehicle insurer Progressive uses machine-learning analysis of the data collected from cars to price its usage-based insurance premiums accurately. This results in better management of underwriting risk. Meanwhile, Las Vegas turned to a machine-learning solution to secure its smart city initiative. The goal for the AI here is to detect and respond to threats automatically in real time.

Implications for Enterprises

For many enterprises across all industries, artificial intelligence has the potential to boost the value established by the deployment of IoT. This enables better offerings and operations to the enterprise, which gives them a competitive edge regarding business performance.

Moreover, many executives who are considering new IoT-based projects should know that machine-learning for predictive capabilities is now integrated into most major horizontal (or general-purpose) and industrial IoT platforms including IBM Watson IoT, Amazon AWS IoT, PTC ThingWorx, GE Predix, and Microsoft Azure IoT.

There is a growing number of turnkey, bundled, or even vertical IoT solutions that utilize AI technologies such as machine-learning. For example, looking at connected-case uses, BMW's CarData platform allows vehicle owners to share data that can then be accessed by IBM's Watson to IoT. Regarding retail and consumer products, a large number of solutions for replenishment automation and optimization utilize machine-learning to predict demand and allow for a more optimized inventory level for production. Moreover, providers of telematics solutions for the insurance industry for autos are also integrating machine-learning to create a more accurate risk model and better predict claims behavior.

It is possible to use AI technology to get more

value from the deployment of IoT which were initially not designed to be used with AI. For example, a Hungarian oil and gas company used machine-learning technology to sensor data which was being collected during diesel fuel production. Here, the analysis allows the company to predict the fuel's sulfur content more accurately, not to mention that AI also helped them identify some improvements they could make during the production process. Here, the company saved more than $600,000 a year thanks to AI technology coupled with IoT. Many horizontal and industrial IoT platforms which many enterprises might be using are also offering new AI-based capabilities that help them boost the value of their existing deployments.

The Future of IoT is AI

In the future or the near future, it might be rare to find an IoT deployment that does not involve the use of AI in some way, shape, or form. The International Data Corp predicted that AI will support all effective IoT efforts by 2019. Moreover, without AI, the data collected from all IoT deployment will have limited value. There are a growing number of IoT vendors out there that are offering at least basic AI support. Some of the companies that do take the first steps in integrating AI into their IoT deployment are currently reaping the benefits.

At this point, IoT can fall into the general pit of buzzword-vagueness. AI also fell into the same trap, especially when there were new terms such as machine-learning, deep learning, genetic algorithms, and many more. While it is true that many people have at least a vague understanding of IoT or AI, few know what they are. AI is making quick strides, and its development has never been stronger. IoT has been here for quite some time, but the introduction of AI has given IoT the upgrade it needs. Many technology executives and researchers agree that AI will be necessary, regarding functionality, to wield a large number of connected devices online. Its importance is even greater when it comes to the need for making sense of what seems to be an endless stream of data flowing from all the connected devices.

Chapter 6: Opportunities for Artificial Intelligence

Many early adopters of AI and cognitive technologies are reporting that there can be great opportunities for economic gains and job creation. According to a study conducted by Deloitte just last year. There is an overall sentiment for organizations that were interviewed by Deloitte. They believe that the role of cognitive technologies plays an important role in the overall operation processes in their organization. About two-thirds of US executives who are aware of such technologies said that they have training programs for employees to learn how to develop their own cognitive technologies or find a solution to work alongside them. However, unlike what many people fear, the respondents for the study did not see job loss as a consequence of their collective AI-related efforts. Only 69% of them believe that there will be minimal to no job loss within the next three years.

In fact, the report also said that more than 25% of the organizations see that newer jobs will arise with the adoption of AI and cognitive technologies. When the respondents were asked about the perceived benefits of AI and technologies, many of them put workforce reduction at the lowest. Their main interest is the enhancement of their own products or services, with better decision-making, optimization of internal business operations, and the ability to create new products trailing close behind.

The CEO of Deloitte, Cathy Engelbert, believe that instead of being concerned about the rise of the machines which will replace humans, everyone should strive to find ways to work in unison with them. The ability to use new technologies to revolutionize the workforce will ultimately lead to new, greater, and more exciting opportunities to build high-value skills for workers. According to the same study, most of the organizations surveyed believe that cognitive technologies play an important role in their internal business processes, and four-fifths of them believe that this technology helps improve their products and services.

Overall, most of the organizations believe AI and cognitive technologies will transform their organizations substantially. Moreover, the study also found that more than a third of the organizations have invested roughly at least $5 million in AI and cognitive technologies. Those investments are heavily leaning toward IT, product development or research and development, as well as customer service. About 75% of the early adopters are currently exploring mature cognitive technologies such as RPA, or Robotic Process Automation, and 70% of them are looking into machine-learning. Half of them utilize deep learning neural networks to train their bots.

The result can be expected. Most of the respondents reported moderate to substantial economic benefits from the use of AI and cognitive technologies. They said they feel that cognitive tools should be used for transformational change as opposed to incremental improvements. Another group, however, believe that they should start small first and see where the technology develops before putting in everything.

Nowadays, all industries are very competitive. Therefore, a competitive advantage is always desired. As such, those organizations are now thinking of how, when, and why should human and machine work in unison to achieve the best possible outcomes. Of course, AI and cognitive technologies are the greatest tools an organization can ever have. However, knowing how to use that tool is an entirely different story. It is worth noting that the implementation of these technologies will disrupt the internal process, decision-making, customer service, and much more. It will take some time for workers to adjust as well, so the true value organizations can get from the use of AI is through its clever use within the context of a company's business, marketplace, corporate culture, and industry.

Chapter 7: Threats of Artificial Intelligence

Before, hunting and gathering were the only means of survival for humans. However, humans are as intelligent as they are lazy. So, they created tools to make their work easier. They started out with sticks and plows and tractors. Agriculture started off as a job that required almost an entire town, to a family, to a few people. The harvest, though fewer people were working on the farm, just kept increasing instead. Abundance was still there, though, so few worked in the field. Of course, technology does not change agriculture alone. It changes everything. Many tools are invented to ease physical labors of all kinds, and they can be called mechanical muscles. They are stronger, more reliable, and definitely never tire, let alone ask for a raise. They just need sufficient energy and their parts properly maintained. Because of these traits, these mechanical muscles are replacing humans in factories. That is good, though. Using mechanical muscles to replace human labor allows them to specialize, even if it means doing other manual labor jobs. Economy and standard of living are getting better because of this. These mechanical muscles take away the hard physical labor so people can use their strengths to work in mental labor instead. When that gets boring, people invent mechanical minds.

Some people may think that humanity has been through this before, but this is a new situation altogether. A lot different from the revolution brought about by the mechanical muscles. When automation is mentioned, people tend to think of giant, custom-built, expensive, yet efficient robots that can only do one job right. They are only good in certain situations. A robot in a car assembly cannot put together an iPhone and vice versa. These robots are already considered to be ancient. There was a new kind of robot introduced in the last decade that raises some concerns.

General-Purpose AI

The perfect example is Baxter who is an industrial robot built by Rethink Robotics.

Baxter was introduced in late 2011 and was succeeded by Sawyer, which is also a robot. These robots are equipped with arms and an automated face. But the most notable feature of them all is how they work.

Baxter is able to see and learn from users by watching them perform the action before it imitates the same thing. He is better simply because he is cheaper to operate and he can do a lot of things as long as the things needed are within his reach. One could say that Baxter is a general-purpose robot, which is a big deal. Think of computers. They initially started out as highly custom and highly expensive, not to mention that they were bulky and pretty much immobile. Then, one day, general-purpose and relatively cheap computers hit the market, and sold like hotcakes. They quickly became vital to everything from communicating to the production of documents, images, services, etc. Now, it is impossible to live a good life without the help of a computer. Such a bot can calculate change or assign seats on an airplane, as well as perform other tasks relatively easily just by installing different software. Because of their versatility, there has been a huge demand for computers of all kinds that make these computers more powerful and cheaper every year.

Here, Baxter can be said to be the computer from the 80s. While his appearance is not revolutionary, his presence is the start of something bigger. It is true that Baxter is operating at a sub-optimal speed compared to human workers, but he only consumes electricity, which is massively cheaper than paying humans monthly. A tenth the speed is still better when it is a hundredth the price. Plus, should Baxter be utilized, there will be tens of him in the workplace to compensate for the work speed while still maintaining financial efficiency. It is true that Baxter is not very smart, but he is smart enough to take over many low-skill jobs. There have been machines dumber than Baxter replace jobs. Supermarkets used to have 30 humans. Now, some of them have only 1 who oversees 30 cashier robots. Amazon opened its first automated grocery store Amazon Go in Seattle that removes the checkout lines and cash registers altogether. Purchases are tracked via cameras and sensors, and payments are processed digitally. Baristas are not safe, either. There are hundreds of thousands of baristas all over the world, and there is a barista robot coming up. Sure, some people may prefer their coffee just perfect, and they'd not trust anyone else aside from their favorite barista, but most people just want to get a decent cup of coffee, which a bot can easily do consistently. Plus, these barista robots are actually a giant network of robots that recognize different people and brew their coffee just how they love it no matter where they are, which is very convenient. Technological changes are often thought of as the fancy new expensive stuff, but the real change comes from the last decade's stuff getting cheaper and faster. They can easily out-compete humans for jobs because they are fully capable of making decisions, which makes them even more frightening than their predecessors – the mechanical muscles.

White-Collar AI

Remember that robots replace physical labor so humans can specialize? Robots are now pushing humans out of these specialized occupations as well. Think of two horses back in the early 1900s discussing the advancement of technology. One is worried because technology will eventually make a horse's works redundant. The other horse said that technology, so far, has only made the lives of horses easier. Horses are not needed to do farm work by dragging that plow all over the fields. Horses are not needed to deliver letters by running from coast-to-coast. Horses are not ridden into battles. All the previous jobs, from farming to mail delivery to battle, are all horrible jobs for horses. Now, technology takes away all those horrible jobs and allows the horses to work in the city where there's a high demand for transportation. For horses, the city jobs are pretty comfortable. The second horse predicts that there will be more jobs for horses than ever. Suppose for a moment, the second horse said that cars would be widely available, then there would still be better jobs for horses that they can imagine. However, humans know better. What happened was the complete opposite from the optimistic view of the second horse. Horses are still here, and they are still working, but it is not what the second horse imagined. The number of horses peaked around the early 1900s, but since then the number has plummeted. To say that better technology raises the number of better jobs for horses is silly. But when horses are swapped for humans, as in "Better technology raises the number of better jobs for humans," suddenly many people think it sounds about right.

Mechanical muscles ended horse jobs in transportation. There is no proof stating that mechanical minds will do exactly that to humans in terms of mental labor. It will not happen immediately. It will not happen everywhere. Still, it will happen in large enough numbers, and soon enough that is going to be a huge problem if humans are not prepared. Unfortunately, many people are not prepared. Some people will still maintain the view that robots cannot possibly replace their jobs by looking at the state of technology now, and its development trend – making life easier for humans. The evolution of biology cannot match the rapid pace of the advancement of technology. Cars ended many horses' careers. The same should be expected for humans with AI. Self-driving cars are no longer the future because they are here and they function as intended.

There are self-driving cars going hundreds of miles up and down the California coast and even through cities where there is heavy traffic. They accomplished this without human intervention, which speaks volumes of how humans will be replaced. The Darwinism theory for machine-learning also applies here. Again, bots do not need to be more perfect than humans. They just need to be at least slightly better. One can easily see how that can be a reality. 40,000 deaths a year are caused by traffic accidents in the United States. Since self-driving cars don't blink, don't text while driving, don't get sleepy or silly, it's easy to see how they are better than humans because they already are. Sure, they may still have some technical flaws, and that traffic accident may not be eliminated completely, but they work, and the number of deaths associated with traffic accidents will be reduced, and that is good enough.

Looking at self-driving cars, calling them so is comparable to calling the first cars mechanical horses. It can even be limiting to call cars as they are because they are capable of doing so much more than horses, and the same could be applied to self-driving cars. They are actually autos: the transportation solution capable of getting itself and the object it is carrying from point A to point B without human intervention. Now, some may see how these autos can replace humans in more ways than one.

Traditional cars happen to be human-sized because they are meant to transport humans. However, tiny autos can work in warehouses, and gigantic autos can work in pit mines. The job of moving things around already covers a massive number of jobs. In the United States alone, the transportation industry has about 3 million people. Extrapolating world-wide, that is roughly 70 million jobs at a minimum. These jobs are over.

Some may argue that unions will prevent it. However, history is already filled with workers who fought the technology that would replace them. Many people nowadays know what happens: the workers always lose. The sad truth is that economics always wins and, as it stands, the incentives for using bots to replace human workers have never been bigger, especially when it come to the use of autos. For many transportation companies, humans are already about a third of their total expenditure. That's just the salary costs alone. For a long haul, it can take a few days for the goods to arrive at their destination. Along the way, the human driver sometimes takes a break to sleep, which costs time and money. Accidents also cost money. Carelessness costs money. Some may think that insurance companies will be against the idea of autos. But it is actually the opposite.

Insurance companies make a profit, not from the drivers who get into accidents, but the majority of those who do not. The autos are coming, and they are the first place where most people will really see robots changing society. Even more worryingly, there are many other places in the economy where the same thing is happening, just less visible. If bots can revolutionize transportation, it will do the same to everything else.

It is easy to look at Autos and Baxters and think that technology has always gotten rid of low-skill jobs people do not want to do anyway. Such a replacement allows people to specialize in better jobs, as they have done. However, even if the challenges of getting millions of people to receive higher education for a better job, white-collar work is not spared from bots either. If the job requires a person to sit in front of a screen, typing and clicking, then the robots will eventually be able to do just that as well. There is no need to build a robot with physical hands to type and click. Why not just have a robot installed right in the computer to do work? Software bots are intangible and can work at a blistering speed. From a company's perspective, white-collar workers are costly and numerous, so the incentive to automate their work is greater than even low-skilled work.

Of course, software bots need programmers

behind them first. Unfortunately, that is a job for automation engineers. These are highly-skilled programmers whose job is to replace white-collar workers with a software bot. While it may seem an impossible task even for the world's smartest automation engineer to make a bot do white collar jobs, and those who think so are probably right, it is worth noting that technology thus far has accomplished what was initially thought impossible. Humans have achieved aviation in a vehicle a lot heavier than air, seafaring on a vessel so large yet able to float, or even soar above the skies to a celestial body. The cutting edge of programming here is not about super-smart programmers writing bots. It is about programmers developing bots that can teach themselves to perform tasks that their creator cannot teach them to do.

The bottom line is that bots can learn how to do things even when it receives little to no instruction or data. They can still learn as demonstrated by OpenAI's bots that can thwart the world's top eSports players in 2 weeks of machine-learning. For example, the stock market hardly involves humans nowadays. It is mostly bots that taught themselves to trade stocks and trading stocks with other bots that taught themselves. Because of that, one can no longer find traders in New York Stock exchange. All that can be found is a big screen. In fact, there are bots out there that can write Harry Potter books, amongst other things. Some newspapers published nowadays are written by a bot, and many people cannot tell the difference. While this book is written by a human, future books may not. Again, the bots need not be a great author. They just need to do a slightly better job than most humans, and it is not a really difficult thing to accomplish. Some companies started teaching bots to produce content such as news, sports, and even reports. A lot of human work such as paperwork, writing, or decision-making are easy prey for the bots, not only because they can be automated, but also because there are a lot of people working in this field, which makes the replacement of humans with bots a huge incentive. But what of other professions?

When "lawyer" is mentioned, people tend to think of trials. However, a lawyer's work is mostly preparing legal documents and predicting the outcome and impact of lawsuits, and a small portion of their work is in the courtroom. There is also a conveniently bot-friendly job called "discovery." Here, the lawyer needs to go through all the documents with which they use to find the clue or fact they need to win the case.

This sounds a lot like a job that bots can perform very well. As such, there are already not a lot of human jobs in many law firms because bots can go through hundreds of papers in a very short time without making mistakes. These research bots have already crushed their human competitors by a long shot, not only regarding cost and time but also accuracy. Humans can get sleepy and overlook that one out-of-place transaction or that one email amongst a million. Bots don't get sleepy when they plow through a million emails.

To make matters worse for humans, that is only the simple stuff. Back in 2011, Watson was created by IBM. He was designed to be a computer system that could answer questions. Developed by DeepQA from the same company, which is led by David Ferrucci, the principal investigator, Watson was named after IBM's first CEO, Thomas J. Watson, an industrialist. Watson was initially developed to participate in a quiz show known as "Jeopardy!" Some people may have seen him in action during that game show. Watson literally destroyed his human competitors with ease. One may argue that Watson may have access to more information than his competitors and be able to memorize it better, which is why he won. It is true, but at the same time, that is why a bot is better than humans. Plus, Watson is not designed for quiz shows. That is only a small side-project for Watson. His actual job is being a very good, if not the best, doctor in the world by understanding how patients describe their own sickness in their own words and come up with an accurate diagnosis. Watson is already in full operational deployment back in Slone-Kettering, and he is giving helpful guidance on treatments for lung cancer. Again, Watson is not, nor does he need to, be perfect at what he does. There is and will be a time when he makes mistakes, but at least it will occur less frequently compared to human doctors. To think that human doctors, no matter how much experience they have, are perfect would be unrealistic. They sometimes make mistakes, and fatal ones at that. Moreover, the patient's medical history is a massive puzzle to be solved in and of itself. The best treatment requires a review of the patient's medical history, and a full understanding of each and every drug and how it interacts with other drugs. This alone is already beyond the scope of a human's capacity. There are also research bots that test hundreds of new drugs simultaneously. Human doctors can improve through individual experience. Doctor bots can do the same, except that they are fully capable of sharing their experience with other bots. Think of them as a large group of bots with a single mind. There are also research robots that test thousands of medicine at a time. Such a capacity is beyond a human's physical limit.

Plus, doctor bots can remain up-to-date with the latest medical information disseminated across the globe. They can track the conditions of their patients no matter where they are and make correlations, which is beyond what a human can do. Of course, not all human doctors will go away, but when one has access to a highly-skilled doctor bot that is comparable, if not better, than its human counterpart and that they're as far away as a phone call, the need for human doctors will be less. As such, both white and blue collar workers probably need to start worrying about the introduction of AI and automation. But there is yet another field of profession that some may view as the last hope of humanity's occupation – creativity. Sadly, artists and musicians are not really safe from bots.

Creative AI

While creativity may feel like magic, where people can turn their thoughts into words or pictures that evoke emotions in others, it isn't. It is true that the brain is a complicated machine, arguably the most complicated machine in the whole universe, but it is still a machine that can be simulated. This fact alone has not stopped humanity from trying to do so. There is a belief that, just like mechanical muscles pushing people into doing mental work, mechanical work would do the same, except that people would move to a creative occupation. But even if one is to assume that the human mind is magically creative (which it is not, but for the sake of argument), artistic creativity is not really a reliable source of income because not many people make a lot from it, not to mention that they do not even make up a noticeable portion of the workforce. Since successful creative individuals rely on being famous, making a living out of creative work is not ideal. It is ludicrous to think that there can be an economy based on poems and paintings.

When it comes to creativity, there is already one bot out there. Emily Howell is a bot that can produce music, free of charge. Many people, when put to a blind test, cannot tell the difference between her work and a human's work. Creative bots are also on the rise. Chess was initially thought to be a human-only sport and that bots can never be as good as humans until they are proven wrong.

This is a reality that many people will find hard to accept, and many people will reject it. It is easy to be cynical of the endless and idiotic predictions of futures that never are. That is why it is important to emphasize again that bots, and how they can replace humans, is not science fiction. These robots are here right now, and their capability is already demonstrated in many fields, be it a blue-collar job, a white-collar job, or even a creativity-based profession. The robot revolution is different from the previous economic revolutions that humans have been through. Horses did not lose their jobs because they became lazy and did not want to work. They are just unemployable because there are other means of travel that are cheaper, better, and more convenient. Now, there are not a lot of jobs for horses, let alone the ones that yield a decent profit consistently. The same could also be said for humans. Many talented individuals will remain jobless just like the horses, and it is not even their own fault. Some people may think that there should be new jobs popping up that are only exclusive to humans. There is also another matter to think of. According to the US census in 1776, there were only tens of kinds of jobs available. There are hundreds of jobs, but the new ones do not make up even half of the existing workforce. Presently, the top three jobs are transportation, retail salespersons, and first-line supervisors, which amount to more than 3 million. Cashiers, secretaries, managers, sales representatives, registered nurses, elementary school teachers, janitors, and cleaners follow suit respectively, with their number close to 3 million. All these jobs have been here for hundreds of years, and they can and will be automated.

During the great depression, the unemployment rate was 25%, which made up about half of the current workforce. Given the fact some machines can do most of all the top jobs of millions of workers worldwide, this is a huge problem. Of course, this book is not meant to frighten readers about the inevitable fact that robots will replace every human on earth. It is just as a reminder that automation will happen eventually because this has always been the trend in technology – being a tool to produce abundance for little effort. Humanity as a species needs to start thinking now about what do when large sections of the population are unemployable, especially when it happens not because of their own fault. What can fresh, perfectly capable college graduates do in the future where, for most jobs, humans need not apply?

Before further plans of action can be formulated, it is important to know how these bots can be used across industries. Only then can the severity of automation be accurately gauged.

Chapter 8: How AI is Used in Different Industries

Come to think of it, many people should feel at least grateful that they do not have to spend most of their lives doing their jobs manually. As of now, everyone lives in a time where a lot of work is taken over by machines, software, and other automatic processes. There is no need to spend hours building a car with ten people when a bunch of robotic arms can do the same in less time with greater accuracy. There is no need for an interpreter when Google Translate is available. As such, AI has a special place in every single technological advancement today. AI is simply here to assist its humans in doing regular tasks.

As a consequence, modern life has also become more advanced with the use of this technology. So, how are they important? How are they implemented?

A Great Help for Humans

At its current stage, it is safe to say that AI systems are good enough to help reduce human efforts in many areas. This frees the humans so they can put their efforts where they really are needed. To carry out different activities in the industry, many companies use AI to create machine workers that can do the same jobs as humans, although the bots are a lot faster and provide arguably greater or more accurate results. AI is here to make the world error-free and efficient. Recently, many sectors have started using AI to cut down on human efforts, not to mention increase efficiency and productivity.

Finance and Banking

When online transactions have rapid growth every year, the finance and banking industry has to deal with major problems such as identity theft and fraud loss cases. Here, AI can bring the financial industry the much-needed cybersecurity to the next level by utilizing deep learning technologies to analyze patterns and spot suspicious behavior and prevent potential fraud altogether.

An example of such a bot in action is from PayPal, which managed to reduce its fraud rate to below 0.5% of revenue by utilizing a sophisticated deep learning system that is capable of analyzing transactions going on in real time. Moreover, AI can also assist in the mundane tasks in the financial industry by assessing credit quality or automating client interactions, which saves a lot of time and money.

Heavy Industries

AI is used in many production units in many big manufacturing companies. These AI systems are used to form an object to a specified shape, move said object from point A to point B, heat it up, etc. This application is utilized in many companies to get a lot of jobs done efficiently on time. It is also used to keep records of all employees and the important data of the company which is being stored, which would then be extracted whenever needed. Heavy industries thrive greatly on the AI system because the tasks in those industries are the easiest to automate, allowing the companies to save a lot of money in operating costs.

Healthcare

In the healthcare sector, in addition to Molly, AI can also provide invaluable assistance in the form of analysis of complex medical data such as X-rays, CT scans, and different screenings and tests. By using the patient's data and knowledge sources outside such as clinical research, medical professionals are capable of building a personalized treatment path for virtually everyone.

For example, Babylon AI doctor app utilizes speech recognition to consult with patients, then checks their symptoms against a database, and then gives them the correct treatment. Microsoft's Handover project also utilizes machine-learning to make predictions about the most effective drug treatment for cancer individually.

Retail

Retail is perhaps one of the easiest targets for AI and automation. Its application seems to be designed to replace retail entirely. Of course, it is more than just conversation intelligence which allows companies to talk to their customers, or follow leads through the analysis and segmentation of sales calls using speech recognition and natural language processing; there are also chatbots and virtual customer assistants to worry about. Combined, these bots are enough to allow retail companies to run and provide customer service 24/7, not to mention that they can answer basic questions without human intervention.

There are recommendation engines that use machine-learning technology to predict and analyze customers so their shopping experience can be personalized. Different people will see different recommendations. Major e-commerce platforms such as Amazon rely heavily on the use of recommender systems. As a result, their revenue skyrockets (by about a third). There are also geo-targeted sales campaigns, catered to different people in different geographical regions. These campaigns use price optimization to produce just the right offer, and Darwin Pricing's dynamic pricing software is the perfect example. These pricing optimization systems also use machine-learning technology extensively. Darwin Pricing, in particular, uses artificial neural networks to model price expectations for different places. That way, retailers can offer effective discounts to boost sales without compromising their profit that much.

Technology

Unlike popular belief, technology companies do not primarily build AI, and to an extent, replace other people's jobs. Artificial intelligence also has its use in technology companies. In fact, tech giants such as Google, IBM, or Apple are known to grab smaller AI companies to do their work so they have a competitive advantage.

When people have technical problems with their devices, the diagnosis and fix can be hard to understand for normal people. That is why the ability to understand what the customer is saying and knowing exactly what is wrong with the device is indispensable. Here, chatbots or virtual customer assistants that use speech recognition and natural language processing fit the role perfectly.

Other than chatbots that are used by small- or medium-sized enterprises, market leaders also need to build their own intelligent voice assistants. There are Google Home, Microsoft's Cortana, or Apple's Siri that serve as a personal assistant. They are fully capable of analyzing human language and providing appropriate answers.

Translation engines that are powered by artificial intelligence are also a big thing as well, mainly because they revolutionize communication. Not only that people can gain instant access to a handy translator online such as Skype which offers real-time AI translations. Most notably, Google Translate, initially had its capacity limited to the words typed into the box, now has mobile applications that enable users to take a photograph of any text they find and translate it by scanning the image. This can only be accomplished using machine-learning or even deep learning to achieve image and letter recognition, enabling instant translation between languages.

Facebook also utilizes face and image recognition. Users now are informed when an image of them is uploaded and asks whether the user wants to tag themselves in that picture, even if that picture is posted by a complete stranger.

Of course, there are countless uses and implementations of artificial intelligence in the tech industry. Its relevance is still on the rise, and it does not seem that it will go away any time soon.

Higher Education

Some may think that lecturing a class of tens of students is something that cannot be automated, especially when the students and each individual class are so different that automation is outright impossible. However, artificial intelligence is slowly creeping into this field as well. It is true that, by automation, one is implying that the same process is repeated for all students. However, the bots can do it differently. Instead of approaching the problem of lecturing hundreds of students using the same technique, bots can utilize personalized learning that tailor educational content to the needs of each and every individual student. This level of customization is well beyond a human lecturer's physical and mental capacity. Here, data analytics assist in the implementation of adaptive learning programs by enabling the educators or lecturers to collect and analyze data from students by looking at their performance, learning style, and possibly even lifestyle so the learning programs can be adjusted accordingly to achieve the best possible result.

Oregon State University is already using these adaptive learning technologies for their students who are taking some of the most difficult courses that come with the highest attrition rates. This makes students more eager to learn because the experience is tailored to them while making the course seem less difficult.

Northern Arizona University has also started to implement the same method across its university. As a result, its DFW (D-grades, F-grades, and Withdrawals) have significantly dropped from 23% to 19%, which is a huge success.

Machine-learning can also be utilized to give immediate feedback on students' writing assignments. The University of Michigan utilizes a program called ATA (Automated Text Analysis) which reviews written submissions. It identifies the strengths and weaknesses of every single submission and even gives recommendations to the students as to how they should revise their papers.

Energy and Utilities

At the present moment, artificial intelligence is still in its early stages when it comes to implementation in the energy and utility industry. Still, several companies in this sector have started investing in the new technology. Artificial intelligence and big data deals in this industry went up tenfold in 2017. Here, industry leaders expect artificial intelligence to make energy systems cleaner, better, and more affordable and reliable.

One of the most popular implementations to watch out for is the intelligent energy forecasts, data analytics to manage intermittent renewable generation as well as self-healing digital grids.

Analysis of the patterns in the power grids to locate their vulnerabilities are also expected to be performed by bots. A project led by the Department of Energy's SLAC National Accelerator Laboratory wants to use artificial intelligence to minimize or even prevent electric grid failures altogether by installing an autonomous grid that can respond to disruptive events very quickly.

Transport

Air transport is undeniably one of the most systematic transportation systems. Because of its complexity, air transport cannot thrive without help from AI. Numerous functions in the machines and management processes are actually controlled by AI. This can range from booking flight tickets to automated check-in and passport control. AI application makes air transport a lot more efficient, comfortable, faster, and most importantly, safer for everyone concerned.

Chapter 9: How Humans and AI can Work Together

Artificial intelligence, as it stands, is fully capable of doing many jobs that are once done by humans. It is already established in this book that AI can do many, though unlikely for bots, things such as translation, providing customer services, or even diagnose diseases with an accuracy better than that of human. It is not a mystery that bots are improving daily, at a pace hundreds of thousands of times faster than natural evolution. Such a pace raised several concerns that AI will eventually replace human workers throughout the economy, which has also been discussed. It is true that such an outcome is possible, and yet there is little known about the future. In the interest of keeping the tone of the book light and provide a balancing positive view of the evolution of AI as opposed to the negative one discussed in previous chapters, there are actually some benefits from the use of AI. AI has never been a better digital tool than before, and it will ultimately alter how everyone's work gets done as well as who gets to do that work. Here, there is an opposing belief that perhaps bots will not replace humans in work, but they will assist in everyday jobs, increasing human capabilities to a whole new level.

It is true that many companies use AI to automate whatever they can in the workplace. However, the presence of AI is only for the sole purpose of displacing employees. That could lead to a short-term productivity gain. Research conducted by Harvard Business Review that involves 1,500 companies, has found that those companies achieve the best improvement regarding performance when humans and machines work in unison. How so?

Where Humans Are Better than Bots, and Vice Versa

The fact is that, with enough time spent on machine-learning, bots are fully capable of just about anything that humans can do. This brings about the belief that the future will be bleak with robots flooding the job market, leaving their fleshy competitors in the dust. Against that view, it is worth looking into the matter a bit closer and see how humans can do a better job than AI, and vice versa. That way, one can get a better understanding of how the workforce is going to change, and how bots are not meant to replace humans, but rather to complement them in their work.

What AI Does Best

When it comes to completing repetitive tasks and solving certain problems that involve going through a massive amount of data, it is no surprise to see bots can beat humans with their arms tied behind their backs whilst blindfolded, and that is only a slight exaggeration. Humans get bored and distracted very easily, mainly because their biological cognition is not designed to go through such work. Robots, on the other hand, are not capable of feeling bored or becoming distracted, so crunching a huge amount of data is right up their alley.

There is a saying: "To err is human," which basically means that everyone makes mistakes. It is true when people are assigned to process and evaluate patterns in data that is so large that it could take them days to go through it all. Bots, on the other hand, can do just that accurately and in a shorter period of time. That is why Deep Blue from IBM beat Gary Kasparov at chess in 1997. That is why DeepMind from Google beat Lee Sedol in the Game of Go just two years ago. Bots use pattern analysis on the data given based on the rules and parameters they are provided. Still, that is pretty much where the potential for AI ends.

What Human Does Best

The very precision of AI is both its strength and weakness. It is true that AI can produce a reasonably accurate output after going through a large dataset, but they are not particularly equipped to deal with ambiguity and gray areas. They perform their jobs based on what they are told, without understanding much of the context or nuance. As such, they are not good at making judgment calls, which have raised many ethical concerns about robots making decisions. That is where humans come in. Humans are a lot better at making an accurate decision.

A little secret about AI and the big data that they are so good at interpreting is the fact that humans are actually the ones who sort, organize, cleanse, and prepare the data for the bots.

Humans are just better at it.

When the economy becomes more digitalized and automated, there will be a need for humans with critical thinking skills. Those are not just white collar jobs either. It would be too inefficient to design robots and AI to fix plumbing or even build skyscrapers. Instead, bots will do as much as providing the data that can help make that work faster, safer, and more efficient.

Verdict

As such, intelligence from human and AI complements each other very well. Both have the qualities that create the perfect worker: leadership, teamwork, creativity, social skills, speed, scalability, and quantitative capabilities. Things that come naturally to people are beyond a machine's capacity, such as making a joke. What is straightforward such as analyzing terabytes of data is well beyond a human's ability. A business requires both, and this shows just how well humans and robots can work together.

To get the best out of this collaboration,

companies need to understand how humans can augment machines in the best way possible, just as well as how machines can best enhance the former's abilities, and finally think of how to redesign the business processes to facilitate this collaboration.

Humans Helping Machines

Here, humans need to perform three critical roles to best complement robots. To get the automation capability, which bots are praised for, there is a need to train machines to carry out those tasks. Then, humans also need to explain the outcome of those tasks, which is important when it comes to results that are counterintuitive or even controversial. Finally, they need to sustain the responsible use of those machines.

Training

Before automation can begin, robots need to learn how to perform that task first. To achieve that, a lot of datasets for training the bots are needed. Translation bots need a lot of data to translate idiomatic expressions accurately. Medical bots need the same to diagnose a disease properly. The same can also be said in financial decision making. Plus, AI systems need to be trained in a way that they can best work with humans, not just regarding bots working with their fleshy colleagues, but also regarding compatibility with end users. Organizations across numerous sectors are currently in the early stage of filling up trainer roles, but the tech giants and research groups already have mature training staff and expertise.

Think of Cortana, the AI assistant from Microsoft. The bot required a lot of intensive training to develop a personality that is just right for the users: confident, caring, and helpful without being bossy. Achieving those personalities for a bot is harder than it looks. It takes a team of poets, novelists, and playwrights to create an AI with such a character. The same also applies to Apple's Siri and Amazon's Alexa. They are designed to reflect their company's brand. For example, Siri has a bit of a sassiness, which many people can expect from Apple.

Now, AI assistants are also being trained to display more complex yet subtle human traits, one of which is sympathy. For example, the start-up Koko, which is an offshoot of the MIT Media Lab, has created a helpful technology that assists AI to be able to sympathize with users. Suppose that a user has a bad day, then the Koko system kicks in and, instead of giving a cold response such as, "I'm so sorry to hear that," the bot will ask its user to talk to it and tell it what's wrong. Then, the bot gives advice to help the person cope with the issue by seeing it in a different light. If the user is feeling stressed, then the bot will give recommendations to change the way of thinking by explaining that the stress can be served as a positive emotion that could be utilized to do good things instead of lamenting on a past situation.

Explaining

When AI reaches conclusions through opaque processes, they need human experts in the field to explain their behaviors to those who do not understand. Those who explain are important when in industries that are based on evidence such as law and medicine. Here, a practitioner needs to understand how AI makes decisions based on the inputs given, and how it weights those inputs individually when it makes a decision about handing down prison sentences or giving medical recommendations. The explainers also help many insurers, and law enforcement understands why that self-driving car took that action that led to an accident or failed to avoid one. Here, it is possible to see explainers becoming an important and well-regulated industry which requires the workers to understand how the bots work and explain to the non-experts. For example, the General Data Protection Regulation from the European Union allows consumers to receive an explanation for why a bot made such a decision such as the rate they are offered on a credit card or mortgage. Here is where the creation of jobs are possible even when AI becomes more widespread. Experts gave an estimation of roughly 75,000 new jobs just to administer the GDPR requirement. Imagine how many similar jobs will be created when bots finally come around.

Maintenance

Companies will need to employ people who can work continually to make sure that the bot is functioning properly, safety, and responsibly. For example, there are safety engineers who are an array of experts who focus on anticipating and preventing harm caused by an AI. Developers of heavy-duty robots that work alongside people on-site are very careful in the development of AI so that the robots may not cause harm to human workers. These people work with analysis from the explainers when AIs cause harm, like when a self-driving car causes or is involved in a traffic accident.

Other groups of these sustainers also ensure that bots uphold ethical norms. When a bot is found to be discriminating against people of color, it falls to these people to investigate and resolve the issue. Similarly, there is also a need for data compliance officers whose responsibility is to ensure that the data fed to the bot complies with GDPR and other consumer-protection rules and regulations. Another potential job could involve making sure that AIs manage information responsibly. Apple, just like many other tech giants, use AI to collect personal information from its users when they use their devices and software. Here, the aim is to make sure that the gathering process brings about the best experience for the user without compromising privacy, making customers angry and tripping the legal alarm. There should be a differential privacy team whose work is to ensure that the AI protects the privacy of individual users as it learns as much as possible about a group of users in a statistical way.

Machines Helping Humans

Here, machines can help humans expand their abilities in three main ways. They can amplify a human's cognitive strengths, interact with customers as well as employees so that workers have more time for more complex tasks, as well as embody human skills beyond their physical capabilities.

Amplifying

AI can boost analytic and decision-making abilities by handing out the right information at the right time. However, that can be used to a certain extent to stimulate creativity. Take Dreamcatcher from Autodesk, for example. The AI is known to be able to enhance imagination, even that of seasoned designers. Here, the designer just has to tell Dreamcatcher a few criteria about the product they want. Perhaps a chair that can withstand 300 pounds in weight, standing at 18 feet off the ground and made using materials that cost less than $75, so on and so forth. Dreamcatcher can also provide information about other works which concepts could be useful. In this case, the AI will point to other chair designs that it finds attractive. The AI then produces thousands of designs that match the criteria provided. When faced with thousands of inspirations, it is not hard to get some ideas to get started. The AI also tells the software which design it likes and does not like, which leads to a new round of designs.

Throughout the iterative process, the AI makes sure that the design it proposes meets the criteria specified by conducting countless calculations. Of course, all designs are not aesthetically pleasing or the concept is something that is not up to the standard that the designer wants. This is where the designer comes in. They just have to use their professional judgment and aesthetic sensibilities to pick out the best design.

Interacting

The collaboration between humans and machines also allows companies to interact with their employees and customers in a more effective way instead of sending a copy-paste response, promising to investigate the matter when both sides know that will not happen any time soon. For example, Cortana is fully capable of facilitating communication between or on behalf of people and transcribe a meeting as well as distribute a voice-searchable version for those who cannot attend. It is worth noting that a bot can provide customer services to a lot of people at the same time and it is possible for the bot to hold down the fort alone and let its human co-workers do other things.

A major Swedish bank, SEB, also uses a virtual assistant known as Aida who interacts with millions of its customers. Aida is fully capable of handling natural-language conversations, and she has access to a vast storage of data which allows her to answer pretty much every question related to the bank. Not only that she can answer questions such as how to open an account or make a cross-border payment, but she can also ask a follow-up question to solve the customer's problems. Plus, she is also tone-sensitive, meaning that she can analyze the customer's tone of voice (angry, frustrated, etc.) and then use the information she gathered to provide better services later. Even if the bot is unable to resolve the problem, the customer's call will be diverted to a human customer service representative. Even so, Aida will monitor that interaction and learn from it so she can resolve similar problems in the future. Here, one can see that the bot will handle basic problems whereas humans will handle more complex ones.

Embodying

Being behind digital entities like Aida or Cortana, AI also has other applications that allow them to embody a robot that works alongside human workers. Equipped with their sensors, actuators, and motors, bots can recognize people and objects, not to mention that they can work with humans safely in warehouses, as well as factories.

In the manufacturing industry, robots are now more than just dangerous "stay clear" robots. They can now be designed to be aware of their surroundings. These context-aware, "Cobots" can handle the repetitive heavy-lifting while a human performs more complicated tasks that need dexterity as well as human judgment such as assembling a gear motor.

At this point, Hyundai is already extending this

"Cobot" concept with exoskeletons, wearable robotic devices that allow its user to perform their jobs with superhuman strength and endurance. Compatibility is also not an issue because these devices adapt to the user and location in real time.

Chapter 10: Ethical Challenges of AI

Nowadays, people are surrounded by AI. It is everywhere, and it is painfully clear that they are here to stay. Most aspects of a human's life is now affected by AI, one way or another. Things such as picking a book or booking a flight, or even determining whether an applicant is a potential candidate for a job, whether that bank loan will be granted, or even whether that cancer treatment should be administered; those decisions are influenced by bots.

All these things and many more can now be determined mostly automatically by complex software systems. Here, the enormous advancements that AI has made in the last few years are almost frightening, but what the AI can do can also improve people's well-being in many ways.

For the past few years, the rise of AI has been inescapable. A lot of money has been thrown at AI start-ups. Most tech companies, such as Amazon, Facebook, and even Microsoft have opened new research labs dedicated to the development of AI. AI has become so well-known that many people generally link that to software – and it is true. It is hard to find software that does not have some sort of AI in it. It is not much of an exaggeration when one says that software means AI.

When it comes to what AI could bring to the table for humanity in the future, opinions are divided. Some said that AI will replace humans almost completely. Some say that AI will create more satisfying jobs. Some who have watched Terminator will say that AI will take over the world. They even went on further to say that because humanity has put AI through torturous machine-learning systems, in which bots need to go through millions of photos or document and put out a satisfying result or be deleted, they could only hope that AI will be merciful to humanity when they finally take over. All these opinions point to a major change, a change that is even larger than the introduction of the Internet. However, some technologists have a different view when it comes to the changes to be expected for humanity as a whole in a world chock-full of wonderful machines. Surprisingly, most of them are not as concerned about AI replacing humans regarding jobs, or even less so about whether AI will take over the world. Instead, their concerns center on the question of ethics.

The challenge now is to ensure that everyone can reap the benefit from this technology. The key issue is trying to work out how to make sure that machine-learning, a data-driven AI technique behind a lot of AI successes, can benefit or improve the society as a whole. It is not just for the people who control society, but for everyone. AI has proven itself to be most useful and effective at practical tasks such as labeling photos to recognizing speech and written natural languages, or even helping to identify diseases. The only problem, of course, is to somehow bring those capacities to everyone.

The biggest problem is that the algorithms are so complex that it is impossible to understand how or why the AI system does what it does. Again, a single line of code can be interpreted, the general purpose of a bunch of them can be vaguely grasped, but the whole is a mystery. All that is known is that AI can do the job thanks to machine-learning. To many people, all they want to know is whether the method works. What goes on under the hood is relatively irrelevant. Here, the challenge is to think of how to monitor or audit how AI fills the many important roles across industries.

There is a problem surrounding computer systems. When they become so complex, it could prevent them from receiving the scrutiny that they need. Ongoing operation without investigation or not even knowing how things work in the first place is dangerous. Doing anything based on trust alone, while it does foster goodwill, is risky from the outset. Everything is needed to be well-understood and scrutinized if one hopes to regulate anything – especially when it comes to AI. Jonathan is concerned about the reduction of human autonomy because their systems which are now assisted by technology, have become more complex than ever and they are tightly coupled. He advised against setting up the system, let it run, and forget about it. There might be problems with how a system evolves one day.

AI require oversight, but at this stage, it is not clear how that could or should be accomplished. At this point, there are no commonly accepted approaches. Without an industry standard for testing these systems, it will be hard for these technologies to be implemented widely.

However, it is clear that the regulatory bodies

often play the catch-up game in this fast-moving world, especially with the latest trend in cryptocurrency. Many companies are already testing and exploring the effectiveness of the use of AI to make decisions about parole or diagnosis of diseases in critical areas such as the criminal justice system and healthcare. However, when one outsources such crucial decision-making to machines, the risk of losing control arises. Who can say that the system makes the right call in any given case? How do they decide who lives or who dies? Again, all that is known is that AI takes in a lot of data and then produces an outcome. What goes on in-between the input and output are not known, and ethical practices rely on knowing the entire process – which is a mystery. This poses serious ethical questions. In the healthcare sector, a patient is not administered the medication he needs because the AI determines that he cannot be saved. The AI can be right in some instances, but what if, through the power of the will to live, that patient can recover only if he gets the medication he needs?

A famous horror movie, Saw, looked into a similar idea. In the film, the antagonist was denied his insurance money for his cancer treatment only because of the company's "policy," which happened to be designed using various variables such as lifestyles, habits, wealth, etc., to deny or grant coverage in certain areas. What the company failed to take into consideration is the man's will to live. When faced with a life-or-death situation, humans can do things that they cannot. They can run faster than they usually can. They can go without food or drink for days. They can survive a fatal injury. There have been cases of those who fully recover from terminal diseases such as AIDS through the use of medication (some of them used traditional, untested medication) and they recovered although the medicine does not guarantee that they will ever recover. These survival instincts, while they do not work consistently, save many people's lives. The antagonist realized this after he attempted suicide by driving off a cliff. He did suffer a terrible injury, and the fall should have killed him, but he miraculously survived. He then went on to test the fabric of humanity. The point here is whether or not the AI will take miracles into consideration. As it stands, it relies solely on certainties and regularities. Therefore, there will be a time when the decision made by the AI will result in the death of a patient when that patient could have survived. Looking at the application of bots in the insurance industry, one can easily see that they will adopt the same method covered in that horror movie. Worse still, unlike in the movie when the method of decision-making is known, the same could not be said for bots. Who knows what the bots look for when it is time for them to make an important decision?

Danah Boyd, the principal researcher at Microsoft Research, said that there are a lot of serious questions about the values that are being coded into such a system, as well as who is ultimately responsible for them. There is an increasing desire by many regulators, civil society, as well as social theorists to see that these bots remain fair and ethical. However, these concepts are fuzzy at best.

One area that is fraught with ethical issues is the workplace. It is established that AI will enable robots to perform increasingly complicated tasks and displace an increasing number of human workers. For instance, a supplier for Apple and Samsung, China's Foxconn Technology Group, has announced that it will replace about 60,000 factory workers with robots. Ford's factory in Cologne, Germany, also puts robots alongside humans in their factory. In many factories, humans already work alongside robots.

Worse still, if the increase in automation has a

major impact on employment, it could result in feelings of being replaced amongst humans, which could have a major impact on their mental health. Ezekiel Emanuel, a bioethicist and former healthcare advisor to Barak Obama, said that the three main things that give meaning to a human's life are: meaningful relationships, passionate interests, and meaningful work. Meaningful work here is a crucial element of a person's identity. He also pointed out the fact that, in regions where jobs have been lost when factories close down, the worker faced an increased risk of suicide, substance abuse, and depression.

Kate Darling, who specializes in law and ethics at the Massachusetts Institute of Technology, said that companies will follow their market incentives. They will always want to get more for less, and that is a common practice and not a bad thing in itself. Still, one should not rely on companies being ethical just for the sake of it. Having regulations in place certainly helps. Societies have done this before when it comes to privacy, or whenever a new technology has been introduced. All that is needed is figuring out how to deal with it. Many major companies, such as Google, already have ethics boards in place to monitor the development and deployment of their own AI. There is also an argument that such an ethical practice should be more common. It is important for humanity as a whole to continue to move forward with the introduction of innovative technologies, but it might get to the point where the creation of structures is needed. While the existence of Google's ethics board is known, what it actually does is not. However, Facebook, Amazon, and Google launched a consortium that hopes to develop different solutions to the minefield related to safety and privacy concerns that AI poses. OpenAI, which is a start-up backed by Elon Musk that created bots that could defeat the best Dota eSport player, is dedicated to the development and promotion of AI which is open-source and for the good of all. It is crucial that machine-learning is researched openly and disseminated through open publications as well as open-source code so that everyone can benefit from the rewards that AI brings.

Ethical Issues

If one seeks to develop the industry with an ethical standard, and hope to get a full understanding of what is truly at stake, then there is a need to create a group of ethicists, technologists, and corporate leaders. It is a question of using AI to make humans better at what they do best. The work toward AI should not concern too much about whether robots will take over the world, but rather how they can be used to assist humans in thinking and making decisions, rather than replacing human altogether. When new technologies become widespread, they often create ethical questions. When AI is implemented into weapons, who should be allowed to own them? When it comes to the press, who should be allowed to publish? When it comes to surveillance drones, where should they be allowed to go? Such questions are raised by those who are genuinely concerned about how bots will develop, those who see the implications hidden beneath all the promise that bots bring. It is up to society as a whole to figure out ways to address these problems through civil, informed discussion to create the best legislation. To that end, several issues need to be addressed.

Unemployment

When it comes to bots, the most immediate

concern for many people is the fact that AI will replace workers in a wide range of industries for both blue and white collar workers. However, AI is not a job killer. Instead, it is a job category killer. When mentioned in the context of occupations, AI brings about conflicting opinions. Both research and experience show that it is inevitable that bots will eventually replace many categories of work, especially those in transportation, retail, government, professional services employment, as well as customer service. Take a look at trucking jobs, for example. In the United States alone, there are millions employed in trucking alone. What happens to them when the self-driving trucks from Tesla become widely available? While millions could lose their jobs, self-driving trucks seem to be a more efficient, and even safer choice when one considers the lowered risk of accidents, let alone the money saved for corporations in the trucking industry. The same could be said for office workers as well as the major workforce in many developed countries.

At the same time, companies will have enough human resource to allocate to a better, higher value task instead. Based on what history has shown, every introduction of a new innovation does not destroy jobs, but rather move them elsewhere, and new job categories are created. The same could be expected from AI.

However, there is a problem. It is true that people will have a lot more time when the bots take away the time-consuming portion of the job. However, many people still rely on selling their time to make enough to sustain themselves and their families. The best thing to hope for here is that the time will be spent on non-labor activities such as caring for families, engaging in communities or learning how to contribute more to society.

Of course, the move to the new age of digital transformation will undoubtedly raise concerns about labor displacement, regardless of whether AI exists or not. What AI really does is speed up the digital transformation across many business processes. When companies look to adapt and implement AI into their processes, it is best for an employee to have an honest and open conversation between employers and employees. This is because, according to experience and numerous research, it is proven that use the augmented intelligence approaches, in which AI assists humans in their jobs rather than completely replacing them, shows a faster and more consistent return on investment for all organizations. Plus, such an approach is welcomed by employees. Who doesn't want their jobs to be easier, especially when employers take away the big data crunching part? People just feel more comfortable working with bots than being replaced by them.

Perhaps, in the future, when the transition to AI-based jobs is complete, people may think that it is even barbaric that human beings were required to sell their time just to make enough to support their family.

Inequality

The economic system is mainly based on compensation for the contribution to the economy, which is assessed mostly using an hourly wage. Here, many companies rely on hourly work when it comes to product or services, mainly because it is a lot more measurable. However, when companies turn to AI, they can produce a lot more within an hour. They can drastically cut down on relying on the human workforce. Therefore less money will go to the human worker. That basically means that the owner of AI-driven companies will make all the money.

Presently, there is already a widening wealth gap. Start-up founders take a large chunk of profit home. Back in 2014, three companies in Detroit made the same amount of revenue as the three largest companies in Silicon Valley. The noticeable difference between the two regions is the fact that the companies in Silicon Valley employed about 10 times fewer employees. Therefore, those who own AI-enabled companies will make a lot more. There should be an equalizing factor so that the wealth gap does not widen too much.

Humanity

As it stands, bots are learning to become better humans. They are getting better at modeling human conversation and relationships. Back in 2015, a bot called Eugene Goostman won the Turing Challenge for the first time by successfully convincing the panel of judges to believe it to be a real boy. The Turing test is more about developing the AI so it can anticipate certain questions from humans so it can pre-form and give semi-intelligible answers, and less about making it intelligent enough to make people think that it is human. Here, the panel of judges was required to use text input to chat with an unknown entity and then they had to guess whether they were talking to a real human or a bot.

This result is merely a milestone for what is to come, as it is expected that people will interact with bots very frequently as if they were humans, whether it is regarding sales or customer service. Humans do not have infinite attention and kindness to bestow upon others, whereas bots can never get tired of building relationships.

Not many people notice this, but everyone has witnessed how robots can trigger the reward center in the human brain, such as click-bait headlines and video games. These clickbait headlines on social media are usually optimized with A/B testing. It is a basic algorithmic optimization for content to capture the reader's attention. This is one of several methods employed to make video and mobile games as addictive as they are. It is said that tech addiction is the new frontier of human dependency.

On the other hand, while they are used to fuel tech addiction, they can also be used to direct human attention to take action for the good of society as a whole. If it falls into the wrong hands, it will do more harm than good.

Biases in Algorithms

AI learns through machine-learning, or deep learning. It requires training data. Ultimately, the data those bots are given will influence how they make their decisions. That means the algorithms can contain biases if the data given is incorrect or contains a certain level of assumption. That way, the bots can reflect, or even magnify the biases that are present in the data.

For instance, suppose that a bot is trained based on a dataset that is sexist or racist. The bot will make predictions based on those sexist or racist data. Certain bots have mislabeled black people as gorillas or even charge Asian Americans higher prices for SAT tutoring. It is not that the bots are meant to be racist, it is just the fact that the data fed to the bots contain those controversial elements that ultimately influence the algorithms – the minds of the bots. Some bots try to avoid problematic variables such as race, but they have trouble disentangle proxies for race such as zip codes. There are bots out there that are trained to determine a person's credit-worthiness or whether that person should be hired. There is a major challenge for these bots because they might not pass the disparate impact test that is used to determine discriminatory practices.

Of course, one way is to determine if there are biases in the minds of those bots, but it is already established that the inner working of the bots is a fiercely guarded secret. Bots are owned by corporations, and they cannot be accessed publicly. It is true that how the bots think cannot be understood, because, at their operational phase, the algorithms will be so complex anyway. However, it is possible to detect biases just by looking at the data that the bots are fed. Even so, corporations will most likely hide that dataset from the public. Here, it is a question of balance between openness and intellectual property.

Transparency of Algorithms

Consider this an extension from the previous ethical concern. It is a problem that companies will not release their bots to the public for scrutiny. It is another problem that certain algorithms are obscure even to their creators, thanks to deep learning.

Deep learning is a growing technique under the machine-learning umbrella that allows bots to make very accurate predictions. However, as explained before, both the bots and their creators can't explain exactly why the bots make such a prediction. This is yet another serious ethical concerns.

Take bots that are used to monitor teachers and fire them, for example. When bots fire teachers, they cannot explain why the teachers are fired. They cannot produce a chart of things to look for to fire a teacher. The best explanation that their human creator can come up with is pointing to the data which the bots are fed.

The problem here is thinking of a way to balance the need for an accurate bot with the need for transparency for the people whose lives are at the mercy of the bot. When humanity is at a crossroad, everyone needs to choose between accuracy and transparency. An example of this is Europe's new General Data Protection Regulation. If it is true that humans are most likely unaware of their own true motives for acting, shouldn't a bot be better at this?

Supremacy of Algorithms

Along the same line, there is a slightly different

concern that comes up from the previous two problems. If bots are trusted to make decisions, who will have the final say in an important matter? Is it going to be humans who will ultimately decide or is it going to be a bot?

For instance, there are already bots out there that are created to determine prison sentences. It is an established fact that the decision of some judges is influenced by their moods. Because of that, some people are sentenced to prison for longer than others simply because the judge was not feeling very well that day. That raises an argument that judges should be replaced with bots instead. However, according to a study from ProPublica, certain bots that are designed to hand down prison sentences are especially biased against black people. To find a "risk score," the bot also considers the acquaintances of the defendant. Of course, a person should not receive a more severe sentence just because he has friends who are also criminals. That would never be accepted as traditional evidence. Yet, the bot takes that into account.

There is also another question about appealing. Should people appeal because the judge is not human? As it stands, the bots are just as, if not more, biased than human judges. When that is the case, then is there truly a fair and impartial decision-making body? When bots are integrated into the judicial system, what would their roles be in the Supreme Court?

Fake News and Fake Videos

Yet another ethical concern comes up about information or misinformation. Here, machine-learning is used to train bots to be able to think of what content to show to different people. Because advertising models are used as a basis for most social media platforms, screen-time is usually used as a measure of success. Because humans are more likely to engage in controversial topics, those stories spread very quickly even though they are biased. Humanity is on the verge of creating viral fake videos that are so realistic that many cannot tell them apart.

For example, it is already established that fake news spreads a lot faster than real news. Fake news is about 80% more likely to be shared than real news. Because of that, many people are trying to influence major political events such as elections or political opinions using fake news. A recent undercover investigation into Cambridge Analytica also caught them on tape boasting about their using fake news to influence elections. It is widely reported that bots had a role in influencing the 2016 US Presidential Election by spreading political propaganda. These automated social media accounts assist in the creation and dissemination of misinformation on the internet in an attempt to manipulate voters and fuel the fire of partisan disagreement.

AI systems are getting better and better at creating fake images, videos, conversations, as well as texts. Even without them, people have enough problem believing everything they see, hear, or read. What happens when the line between a real image and a fabricated image is blurred, when people can no longer tell the truth from lies? Bots can work 24/7 and never tire, let alone make mistakes. As such, they are fully capable of generating a vast amount of fake data in a very short time. This data can be shared amongst users on many social media platforms and essentially flood the platform when they go viral, even though the information is entirely false. Worse still, when they go viral, the news is impossible to stop. Bots can also spread false or heavily altered facts very effectively, therefore amplifying messages and putting thoughts and images into people's heads and altering their opinions about any given matter.

The damage bots can do here is insurmountable. Criminals and state actors can use fake images or audio to cause harm both to a person or to a corporation who would dare to interfere with their interests. Then, all it really takes is a few people to spread those false claims to really change public opinion and swiftly change the public's view.

With the knowledge that news can be fabricated very easily, what will be acceptable as evidence in the courtroom? Is there a way to slow down the spread of false information? Who will decide if the news is true?

Governments and corporations alike should think about how they can withstand the potential damages done by content that is created by AI. In fact, it may be a wise move to consider fake content to be as malicious and dangerous as cybersecurity threats and respond appropriately. Things such as propaganda, disinformation, malicious interference, blackmail, as well as other forms of information crime should be just as harmful as physical attacks or electronic attacks on the systems and organization. As it stands, the world is unprepared for AI being unleashed on unprotected civilians. Corporations who freely traffic in user-generated content should be just as liable as governments to prevent the misuse of AI.

Weapons

AI researchers pointed out that it is possible to create lethal and autonomous weapons systems in less than 10 years. This can be in the form of small drones that can be deployed in the field and eliminate threats without waiting for approval from humans, which is a step more dangerous than the current military drones.

AI researchers have created a video showcasing how small autonomous drones can be utilized to basically commit genocide. Because of this, about 4,000 AI or Robotic researchers have signed an open letter asking for a ban on offensive autonomous weapons. These bots could very well be even more lethal than nuclear weapons. There are several questions associated with the control of bots as autonomous weapons. On what basis should these types of weapons be banned when individual countries want to use it to their advantage, just like nuclear weapons? Even if the ban is possible, what measures should be taken to make sure that there is no secret development of such weapons?

Self-Driving Cars

This is a widely popular concept for the use of bots. Google, Uber, Tesla, and many other car manufacturers are joining in this race to create a fully capable bot that is a better driver than humans. Here, there are also several ethical concerns left unanswered.

Back in March 2018, an Uber self-driving vehicle killed a pedestrian. There was a safety driver mode for emergencies, but its deployment was not quick enough to prevent the accident. When self-driving cars are deployed to the public, who should be responsible when accidents happen? Should it be the company that sells the cars? Should it be the software or hardware engineers who should make sure that the bots are able to respond promptly, and that the hardware is able to respond just as quickly to prevent an accident? Or should it be the driver in the car who should be watching? Suppose that a self-driving car is going too fast because of a software or hardware failure, and the bot must choose between crashing into people or falling off a cliff. What should the car do? This is a widely debated topic, and some have even made it into a small computer game. Also, when self-driving cars are more widely used, and their effectiveness is proven (fewer accidents, fewer deaths) should human driving become illegal?

Rights of the Machines

Expanding from the previous ethical concerns, this may seem to be a silly ethical concern, but when machines become more intelligent and when they are asked to do more and more, how should they be treated? When machines can simulate emotions and act similarly to humans in society, how should they be governed? Should they be considered to be just like humans, equal to an animal, or just an object to be used? To that end, to which degree should they be assigned liability and responsibility when things go wrong compared to the humans that should be in control of them? The traffic accident in March that killed a person sparked outrage, not because it was a traffic accident. It is proven that people die in traffic accidents more than plane crashes. Deaths from traffic accidents are more regular than people think. The reason why people were angry has to do with the fact that there was a machine behind the wheel. Society has not, and may never, accept deaths that are caused by a machine. However, it is unrealistic to expect traffic accidents to be eliminated completely. Still, bots are arguably better drivers than humans when they are fully developed. Traffic accidents caused by self-driving cars will continue to happen, again and again, so the issue of liability and control should be further scrutinized. People need to think of what they can accept and what is ethical and then create laws and regulations to prevent future tragedies. At the very least, fairly assign the degree of responsibility clearly so that the correct individual can be held responsible at the correct degree.

Privacy and Surveillance

When there are security cameras with facial recognition systems, there will be more ethical issues around surveillance. The capability of bots to recognize people just by looking at their face and then tracking them as they move around is concerning, to say the least. Before the development of facial recognition technology, even security cameras didn't really violate a person's privacy mainly because humans were required to keep an eye on the footage at all times, which is pretty much impossible. With bots with facial recognition capabilities, they can look at a large amount of footage much faster at the same time.

For example, China is already starting to use CCTV cameras to monitor the location of its citizens. Some police have also received glasses with facial recognition software installed so they can get information about civilians they see on the street in real-time.

Here, the question would be whether there should be regulations against the use (or misuse) of such technologies? Because social change often starts as challenges to the status quo and civil disobedience, can such a powerful surveillance capacity lead to the loss of liberty and social change?

At the realization that surveillance and facial recognition technology can be so easily abused, Microsoft urged Congress to study it further. According to Bradford Smith, the company's president, the government needs to play an important role in regulating facial recognition technology because everyone is living in a nation of laws. What is striking is the fact that tech giants do not advocate regulations for their innovations regularly. To see Microsoft urging the US Congress to regulate facial recognition speaks volumes of how such a technology can be abused.

Access to AI Technology

There is a saying, "With great power comes great responsibility," It is true that AI can do a lot of good things. At the same time, the bots are also fully capable of doing evil. When the technology becomes more and more powerful, AI can be designed to wreak havoc upon society. What happens when a few evil geniuses decide to use AI to support their malicious actions? Many companies across sectors are already contemplating such a reality and are taking steps to protect themselves from malicious AI attacks. Newer technologies can be used to exploit the vulnerabilities in many systems that are AI-enabled. When those malicious bots get smarter, they can change the nature of their attacks, make them more random and ever adaptive so they can be even more efficient in identifying and targeting the weaknesses in the systems, therefore making their attacks even harder to detect, let alone defend against. This is a terrifying reality, and detecting malicious attacks will only get more and more difficult over time.

Therefore, all organizations that use AI should think of how they design and distribute their AI systems now.

Moreover, Machine Learning service providers, especially those that offer on-demand cloud-based services, should be careful of those who use their services. If malicious users are using their platform to conduct their AI-enabled attacks or other criminal acts, then like financial institutions, governments will take action to crack down on those providers and impose a new form of "Know Your Customers" (KYC) regulations. Those service providers should get ahead of the curve and start their own efforts to ensure that they do not end up on the wrong end of the regulation by knowing who their customers are and their motive of using their platforms.

Error Prevention

Whether it is a human or a machine, intelligence comes only from learning. Here, the systems learn to do what is right based on the inputs they receive. This stage is called the learning phase. When they have been trained sufficiently, they will go to a testing phase in which it is given more examples (unlabeled data) to see how it performs.

Of course, it is virtually impossible to supply the bot with every single possible example out there, so it is ready to deal with the real world. It is true that they can achieve a level of intelligence similar to humans in many ways, but they can still be fooled in ways that humans would not, as illustrated in previous chapters. If humanity relies on AI to bring about the major overhaul in the labor market, security, and efficiency, there should be a way to make sure that those bots perform as planned and that people cannot use it for personal gain.

Singularity

It is a fact that humans are not at the top of the food chain because they can run very fast, have sharp claws, wings, or sharp teeth, or amazing strength. They are up there because of their intelligence and ingenuity. Humans can best beasts that are bigger, stronger, and faster because they create tools to do so. Therefore, one can deduct that knowledge is the key to dominance.

However, this poses a serious question about AI. Will there be a day when they have become so intelligent that they have the same advantage over humans? One may say that the easiest solution is to pull the plug. However, when AI become that intelligent, it will undoubtedly anticipate that move and take measures to defend itself. This is the point known as singularity, where humans are not the most intelligent beings on earth. What then? Bots will eventually come around, so the best way to handle this is by making sure that there are preventative measures so that bots may not take over the world one day.

Ethical AI

Presently, there is a race to become a superpower in AI technology through technological breakthroughs to become the best AI creator there ever was. Several countries such as China, US, Singapore, Japan, and Canada are pumping a lot of money into AI research and development. Behind the hype to continually improve the capacity of AI, there is a need for guidelines and standards that should make the research, design, and use of AI ethical. Now, there is a debate surrounding ethical AI. Whether it can make ethical decisions or is there a need for regulations and what kind.

Challenges of AI

Advancement in AI technology is seen as a benefit for all mankind. However, the belief that bots will inherently do good for society would be too optimistic. Such a view overlooks the critical research and development that are needed to create ethical and safe AI. At this point, there is a lack of transparency for the data flow, and there has yet to be a certification that guarantees AI safety.

There are a few problems associated with the data used by AI through machine-learning. The datasets needed to train the bots are very expensive to collect or purchase because not many people have access to the machine-learning market. Therefore, the data can be biased or full of errors about classes of individuals living in rural areas in low-income countries, or those who chose to not share their data.

What if the bots are trained using good datasets? Suppose for one moment that it is the case, their design or deployment could still encode discrimination in ways such as choosing the incorrect model, building one with discriminatory features. Plus, there is no human oversight and involvement at this stage of machine-learning because the algorithms are so complex to understand anyway, making the system unpredictable and inscrutable. Such a system could also be a result of unchecked or intentional discrimination.

Because engineers do not fully understand how their own bots work, screwed results can be caused by bots that are designed irresponsibly and are fed data that doesn't represent the whole population. This is known as the black box algorithms where the inputs and outputs are known, but the process is a mystery.

What AI did Wrong in the Past

It is possible for AI to act aggressively, according to DeepMind researchers. To prove this, they ran a test to study how AI would react when faced with certain social dilemmas.

Back in 2015, Google Photos labeled two African American people as gorillas. Similarly, Tay, a Microsoft Chatbot, made racist, inflammatory, and political statements just an hour after its launch. This opens up the possibility that biases can be intentionally built into AI that can affect people's lives. To make matters worse, it is difficult to find out if the bot is behaving inappropriately because its algorithms are unknown. Therefore, precautions should be taken before there are long-lasting consequences. For example, if a company does not want to hire women who might become pregnant, they can use machine-learning to find and filter that subset of women out of their hiring process.

At the same time, two multinational insurance companies in Mexico do just that to maximize their profit and efficiency, without paying any heed to any potential implications for the human right to fair access to adequate healthcare. This could be further utilized in the future by gathering data such as shopping history from customers to recognize the patterns associated with high-risk customers and then charge them more, making the poorest and sickest people unable to afford healthcare services.

According to the WEF report that surveyed 745 leaders in business, government, academia, NGOs, as well as international organizations, AI is the technology that has the greatest potential to cause negative consequences over the coming decade.

What Have Been Done to Make AI more Ethical

Two years ago, tech giants such as Google, Facebook, Amazon, IBM, and Microsoft set up an industry-led, non-profit consortium known as "Partnership on AI to Benefit People and Society" to find ethical standards for AI researchers in cooperation with academics and specialists in policy and ethics. This is also a move to calm the public about the potential replacement of bots over humans. Just last year, other companies also joined the party such as Accenture and Mckinsey.

Moreover, many tech companies also take actions to make sure that their technology is protected. DeepMind has set up its own Ethics & Society committee that will conduct research across six key themes such as privacy, transparency, fairness, as well as economic impact including inclusion and equality.

AETHER, an ethics board for AI from Microsoft, also think of things such as new decision algorithms that are developed for the company's services in-cloud. At this point, the board consists only of Microsoft employees, but the company plans to have people from outside to make sure the bots they develop are truly ethical.

State's Involvement

Of course, it is not the corporations alone that understand just how dangerous bots can be if they are allowed to function without any sort of regulation or preventative measures in place. Earlier this year, the Bureau of Indian Standards created a new committee that focuses on standardizing projects that involve cybersecurity, legal, as well as ethical issues. The project can be from any sector, including IT, technological mapping, or using AI for national missions.

UK's House of Lords also released a report to keep the bots in check. The report is titled: "AI in the UK: Ready, Willing and Able?". It calls for the creation of an AI Council. It suggests that the UK government should sponsor more research into AI and convene a global summit in London next year to establish a common framework for ethical development as well as the deployment of AI systems. The Chairman of the House of Lords Select Committee on Artificial Intelligence, Lord Clement-Jones, said that AI does not come without its own risks. Therefore, the adoption of the principles proposed by the Committee will undoubtedly help mitigate some, if not all, the risks. Here, an ethical approach ensures trust in AI technology from the public, and hopefully, the benefits from it can be seen.

The Montreal Declaration on Responsible AI in Canada is also trying to encourage discussions on ethical guidelines. Here, it is noted that AI should ultimately promote the wellbeing of all sentient creatures. The Treasury Board Secretariat of Canada is also looking at the responsible use of AI in the government. Moreover, the Global Affairs Canada also leads a joint effort between universities on AI and human rights.

The European Economic and Social Committee called for a code of ethics that covers the development, the deployment, as well as the use of AI. The goal here is to make sure that what bots do is in line with human dignity, integrity, freedom, cultural and gender diversity, not to mention human rights. There is a need for an approach to AI where bots remain machines that people can control at all times. At the same time, New York City launched a task force to become a global leader that governs how bots make decisions.

Final Thoughts

Humans are becoming victims of these bots. Some are being unfairly tracked, fired, jailed, or even killed just because the bots are biased and inscrutable. Several ethical questions are surrounding the introduction of AI to human society as well. As it stands, humanity as a whole is unprepared for what is about to come. There is a need to find appropriate legislation for AI in these fields. Of course, legislation cannot be formed without a collective opinion, and that opinion does not exist until discussion and conversations are had about the ethical implications surrounding automation in decision-making, amongst other things. Therefore, everyone should start the discussion about bots and create a habit of thinking about the ethical implications whenever a new technology has been developed. That means the creation of ethical AI is a must.

Chapter 11: Conclusion

To conclude, it is inevitable that bots will eventually come about and they will revolutionize the job market. It is true that bots will take over some jobs, but that does not mean that the introduction of AI is the beginning of the end of humanity. The bleak future where bots are the dominant species could only be a possibility unless humans are prepared. Though humans are not prepared, there is plenty of time to become so. To secure a bright future, ethical bots need to be designed, and newer legislations need to be created to control the development of bots. The future remains uncertain, but it is not too late to take action now.

www.ingramcontent.com/pod-product-compliance
Lightning Source LLC
Chambersburg PA
CBHW031420210526
45464CB00005B/1975